Smart Charging and Anti-Idling Systems

Synthesis Lectures on Advances in Automotive Technology

Editor
Amir Khajepour, *University of Waterloo*

The automotive industry has entered a transformational period that will see an unprecedented evolution in the technological capabilities of vehicles. Significant advances in new manufacturing techniques, low-cost sensors, high processing power, and ubiquitous real-time access to information mean that vehicles are rapidly changing and growing in complexity. These new technologies—including the inevitable evolution toward autonomous vehicles—will ultimately deliver substantial benefits to drivers, passengers, and the environment. Synthesis Lectures on Advances in Automotive Technology Series is intended to introduce such new transformational technologies in the automotive industry to its readers.

Smart Charging and Anti-Idling Systems

Yanjun Huang, Soheil Mohagheghi Fard, Milad Khazraee, Hong Wang, and Amir Khajepour

ISBN: 978-3-031-00369-1   paperback
ISBN: 978-3-031-01497-0   ebook
ISBN: 978-3-031-00002-7   hardcover

DOI 10.1007/978-3-031-01497-0

A Publication in the Springer series
SYNTHESIS LECTURES ON ADVANCES IN AUTOMOTIVE TECHNOLOGY

Lecture #4
Series Editor: Amir Khajepour, *University of Waterloo*
Series ISSN
Print 2576-8107   Electronic 2576-8131

Smart Charging and Anti-Idling Systems

Yanjun Huang, Soheil Mohagheghi Fard, Milad Khazraee, Hong Wang, and
Amir Khajepour
University of Waterloo

*SYNTHESIS LECTURES ON ADVANCES IN AUTOMOTIVE TECHNOLOGY
#4*

ABSTRACT

As public attention on energy conservation and emission reduction has increased in recent years, engine idling has become a growing concern due to its low efficiency and high emissions. Service vehicles equipped with auxiliary systems, such as refrigeration, air conditioning, PCs, and electronics, usually have to idle to power them. The number of service vehicles (e.g. public-school-tour buses, delivery-refrigerator trucks, police cars, ambulances, armed vehicles, firefighter vehicles) is increasing significantly with tremendous social development. Therefore, introducing new anti-idling solutions is inevitably vital for controlling energy unsustainability and poor air quality. There are a few books about the idling disadvantages and anti-idling solutions. Most of them are more concerned with different anti-idling technologies and their effects on the society rather than elaborating an anti-idling system design considering different applications and limitations. There is still much room to improve existing anti-idling technologies and products.

In this book, we took a service vehicle, refrigerator truck, as an example to demonstrate the whole process of designing, optimizing, controlling, and developing a smart charging system for the anti-idling purpose. The proposed system cannot only electrify the auxiliary systems to achieve anti-idling, but also utilize the concepts of regenerative braking and optimal charging strategy to arrive at an optimum solution. Necessary tools, algorithms, and methods are illustrated and the benefits of the optimal anti-idling solution are evaluated.

KEYWORDS

anti-idling system, auxiliary-system electrification, powertrain modeling and sizing, working-condition prediction, power management strategies, dynamic programming, AECMS (adaptive equivalent fuel consumption minimization strategy), model predictive control

Contents

CHAPTER 1

Introduction

1.1 MOTIVATION

Different accessory devices (e.g., air-conditioning units) are being equipped in modern vehicles for different requirements. Internal combustion engines (ICE), in most cases, drive these systems even if the vehicle stops, resulting in excessive idling time. The designed efficiencies for large diesel engines result in up to 40% when running on the highway, whereas their efficiencies drop to 1–11% and produce more pollution when idling [1, 2]. Therefore, the progressively increasing requirements on better fuel economy and lower emission levels have driven researchers and manufacturers to develop more efficient as well as less polluted vehicles. Although many types of technologies and products [3] have been proposed recently to eliminate idling, some improvements still can be achieved. As a result, a smart charging system proposed in this book not only meets the power requirements of the auxiliary devices to achieve idling-reduction but also improves overall fuel efficiency.

Figure 1.1 presents the smart charging diagram controlled by a power management unit. According to the auxiliary load of the target vehicle, two configurations of the smart charger can be defined, as shown in Figure 1.1 by employing the alternator, which is directly linked to the vehicle engine through a serpentine belt or to the transmission by using a power take-off (PTO) unit. Therefore, the smart charger can recover a part of the total braking energy. Meanwhile, once the recovered energy is not sufficient to power the auxiliary system, the engine will directly provide the energy in an efficient manner that is ensured by the designed power management strategy (PMS). Thus, these features differentiate it from the current popular products such as auxiliary power units (APUs) or auxiliary battery power systems (ABP). The studied vehicles in this work are service vehicles with air-conditioning/regeneration (A/C-R) units as the main accessory devices, which consume the majority of the auxiliary power. Thus, there is a main power consumption source (i.e., the A/C-R system) and three energy-providing sources (i.e., plug-in energy, regenerative braking energy, and the engine). To meet the requirements of high efficiency and energy saving, the following should occur: (1) the A/C-R system should consume as little fuel as possible while meeting the temperature and other requirements; (2) the recovered kinetic energy during vehicle braking should be at maximum without affecting the drivability; and (3) the aforementioned power sources should be coordinated to pursue the maximum powertrain overall efficiency. Accordingly, the main role of this study is the development of a smart charger to reduce the engine idling. Its components are first sized by a multidisciplinary optimization method and then controlled by the developed power management to guarantee maximum savings on both initial and operating costs. More specifically, by introducing the smart charger to a traditional vehicle, its drivetrain becomes parallel hybrid when an extra battery is added. However, the energy storage system (ESS) in the smart charging system only powers the accessories rather than powering the vehicle, which makes the whole system different from the standard parallel hybrid drivetrain. With the development of the ESS technologies and electric vehicles (EVs) [4–6], the onboard ESS is capable of powering the accessories, e.g., an A/C-R unit, on its own to ensure there are anti-idling benefits. To pursue a low cost and overall high fuel efficiency, the components of the smart charger should be properly sized, which requires a power management strategy (PMS) to decide the power flows among all energy sources and other electrical components.

When it comes to the design of the hybrid powertrains, after selecting the configurations as shown in Figure 1.1, the main challenges are component sizing and developing an efficient PMS to satisfy the desired objective without diminishing vehicle performance [7, 8]. In other words, component sizing and PMS are the two major factors that determine the costs (i.e., initial and operating costs) and the pollution contributions of the hybrid powertrains. Therefore, the results of the component sizing heavily rely on how the components operate and cooperate with each other, which is decided by the PMS. Meanwhile, without appropriate component-sizing methods or results, the PMS cannot optimally coordinate each component for holistic efficiency. In this study, a simultaneous multidisciplinary design optimization (MDO) problem is formulated to solve the component sizing and a rule-based PMS [9]. After the components are properly sized, the real-time sub-optimal PMSs are developed to guarantee the overall high performance in the real-world applications.

Two main types (i.e., rule-based and optimization-based) of the PMS have appeared in literature. To ensure the better control performance, the optimization-based methods are usually designed and used. However, accessing future driving information can assist in reaching optimal performance. Therefore, the driving cycle and the service cycle as the future information are identified in advance. Based on such acquired information, the AECMS, namely the adaptive equivalent fuel consumption minimization strategy, and model predictive control (MPC) strategy are developed for the target delivery trucks in this book. However, the holistic controller, which incorporates the feature that minimizes the A/C-R system power consumption, is also studied (see [3, 10] for more information).

1.2   REVIEW OF ANTI-IDLING PRODUCTS

This chapter reviews the literature on the anti-idling products and technologies. Engine idling when the engine is on and the vehicle does not move contributes greatly to poor air quality. As mentioned before the engine efficiency drops significantly when idling, causing extra energy losses and increased air pollution. As a result, it is highly urgent to decrease or even eliminate engine idling. Much quantitative research has been conducted to show its negative impacts and corresponding bylaws have been issued to completely ban idling in several countries [11–13].

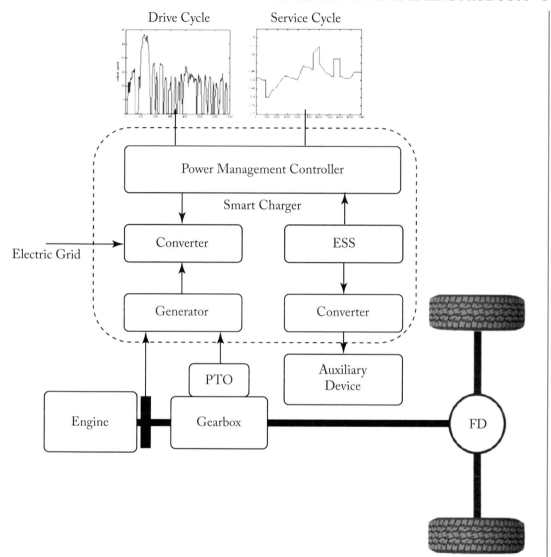

Figure 1.1: Structural diagram of the smart charger.

Numerous devices with specific functions have been proposed by different manufacturers. Generally, according to a comprehensive review of current literature, there are two main classification methods for these technologies. First, they can be classified as the mobile and stationary products, depending on if they move with vehicles. Second, the fully functional and partial functional types are adopted to differentiate products that can/cannot provide all of the power required by auxiliary systems.

1.2.1   MOBILE PRODUCTS

These systems are located in and moving with vehicles, such as automatic engine shutdown devices, APUs, and ABPs. Among them, the last two products are able to independently supply the required power, belonging to the fully functional type products, whereas the remaining refers to the partially functional type [14].

APUs/Generator Sets

APUs, consisting of a relatively small-scale engine and an alternator, are the most popular and conventional anti-idling solutions, which are fully integrated into a vehicle's original HVAC system [3, 15]. A variety of APUs have been proposed by companies, e.g., Thermo King, Dometic, Carrier, Dynasys, Pony Pack, RigMaster Power, and Ecamion. For more information about corresponding APUs, please refer to [16] or the official websites of these companies. APUs usually provide all the required power to turn off the main engine and can thus reduce idling. However, the addition of an extra engine makes trucks more expensive, noisy, and may generate more emissions than the main vehicle engine without appropriately being studied [15].

ABPs

ABPs [17, 18] have recently appeared to be a competitive counterpart to improve APUs. The small engine and alternator in APUs are switched with a set of battery, performing similarly but without the extra noise or emissions produced by APUs. The battery is either charged by stationary type anti-idling devices or by the engine during operation and discharged when the vehicle stops. However, ABPs also face the issues accompanied by batteries (e.g., short lifespan, high costs, etc.) when they are not properly designed. Therefore, other ESSs are presented to replace batteries, such as fuel cell or solar-energy systems, both of them, however, are still too early to be used as they have many other shortcomings [19].

Hybrid Electric Trucks

In 2003, General Motors proposed an electric-diesel hybrid military truck with a fuel-cell APU installed. Hybrid electric light trucks were then introduced by Mercedes-Benz in 2004. Famous for its hybrid commercial vehicles [20], Eaton develops both hydraulic and electric hybrid powertrains. The M2 106 diesel-electric hybrid truck of Freightliner adopts an optional electronic PTO (ePTO) to eliminate idling, making the trucks appropriate for high-idling applications (e.g., tree-trimming). Idling time and fuel consumption are thus reduced by up to 87% and 60% in ePTO mode, respectively. A 5 kW auxiliary power generation unit (APGU) is optional but can bring extra savings [21]. Other truck manufacturers, such as Hino Motors and Mitsubishi Fuso Truck & Bus Corporation [22], are developing their exclusive hybrid technologies and products to reduce engine idling.

Others

Engine management systems also refer to the automatic engine on/off systems allowing truck drivers to program the vehicle to turn engines on/off based on specific parameters—such as a period of time, the engine, or compartment temperature. They can decrease idling to some extent but are not able to power accessories and address the inherent inefficient problem caused by idled engines [11]. In addition, based on research performed by the American Transportation Research Institute, the direct-fired heater is a popular anti-idling product [23], which imports a portion of fuel from the main tank and burns it in a chamber to heat the cabin, whereas Thermal Energy System (TES) [15] is charged when trucks are in operation (especially in high-efficiency periods) during the daytime and adopts a novel cold storage system to supply demanded cooling without starting the engine during the night.

1.2.2 STATIONARY PRODUCTS

Stationary types, referring to truck stop electrification (TSE) systems, are usually located in places where drivers can have access to services including cooling/heating, electricity, and the Internet. They are mainly divided into off-board (single system) and onboard (dual system) categories. Onboard systems (e.g., CabAire LLC and Shorepower Technology [24]) necessitate the installation related devices in trucks, while off-board systems (e.g., Envirodock, AireDock, and IdleAire) offer services via onsite devices [11].

1.3 SUMMARY

Based on the analysis of the existing anti-idling products, to reduce the idling in vehicles with accessories (e.g., A/C-R systems), APUs, and ABPs can clearly achieve this goal by providing enough power for such vehicles. In addition, hybrid technologies are able to eliminate idling as well. However, the optimized smart charger shown in Figure 1.1 still can be a promising alternative approach. Once the components are sized via MDO, the optimized smart charger system will be much more compact, lightweight, and cheaper as it requires fewer modifications to the original powertrains and fewer initial investments. The smart charger cannot only meet the auxiliary power requirements but can also use the regenerative-braking energy to optimize the fuel efficiency. More importantly, compared to the existing anti-idling products, the optimized smart charger exhibits several exclusive features. First, without a small-scaled engine, the smart charger can be much cleaner and quieter than the traditional APUs. Second, similar to an ABP, the smart charger is capable of recovering a portion of braking energy, such that a relatively small battery pack can be sufficient, resulting in a lower initial cost after properly sizing the component. Third, the designed PMS will ensure the entire powertrain is performing in its maximum-efficiency region. Moreover, hybrid trucks are not common due to their high cost, therefore traditional trucks will continue to dominate over the next several decades. Above all, with current designs of hybrid powertrains, the power consumed by the auxiliary devices is usually treated as a constant or even ignored [25, 26], resulting in a not optimal solution. However, the proposed PMS can properly deal with this problem and, thus, can be easily extended to any type of hybrid electric vehicle (HEV).

CHAPTER 2

Powertrain Modeling and Component Sizing for the Smart Charger

To design an acceptable smart charge system, one must consider various factors and criteria. To power the auxiliary devices of the vehicle during idling, the regenerative braking system must be linked into the driveline. The connectivity of the smart charger to vehicle power transformer, total weight, component size, and safety are the most critical factors. It is important that its installation does not make significant changes to the car; otherwise, the new system cost and security concerns will reduce system feasibility for targeted customers. Another vital factor is that the proposed system must be transposable and simple to implement in order to make it feasible for industry. Additionally, the models of the components should be scalable, which can be simply used by the optimizer algorithm [9].

As shown in Figure 2.1, the smart charge system includes various electrical and mechanical parts. The mechanical energy-power input is extracted through the powertrain of the vehicle

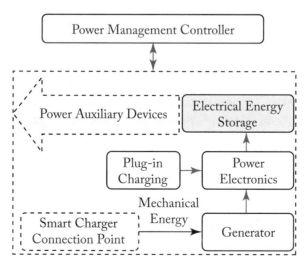

Figure 2.1: Conceptual design of the smart charge system.

from the connection point. This mechanical power is extracted in the form of angular velocity and torque. The potential topology can be divided into two classes. However, the limitation of power transfer and the connection of parts (especially the size of the generator) are critical points for these arrangements. The two major groups of potential configurations are as follows.

1. Serpentine Belt Configuration (Figure 2.2): trucks with relatively low auxiliary-power demand can be designed in such way that the added generator is linked to the engine's belt. In general, the serpentine belt is utilized to power various devices, such as generators, water pumps, air pumps, air compressors, and steering pumps, which are run directly by the engine. In some vehicles, there is a dedicated space for linking another device, which is the additional generator for this study. Therefore, in this configuration, the energy restored is restrained by the characteristics of the serpentine belt and the applicable space for the generator, especially at its maximum stretching capacity. For vehicles with little spare space (e.g., front-wheel drive vehicles), Serpentine Belt Configuration is the most feasible option.

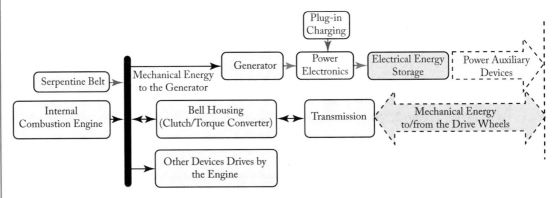

Figure 2.2: Schematic of serpentine-belt configuration.

2. PTO Configuration (Figure 2.3): In this configuration, a PTO module connects to and powers the generator. The power produced by the engine is transmitted to the wheels through the powertrain. Ideally, the power extraction for the generator can happen in six locations. In each of these locations, a different PTO can extract the power. However, during vehicle stops, PTO must let the smart charger extract power from the engine. This situation is possible if the PTO is linked to powertrain at a location between the engine and transmission. Based on a review of different types of PTOs, it was determined that "Transmission Aperture" PTO would allow for this. In most heavy vehicles, as shown in Figure 2.4 (the testbed in the Mechatronics Vehicle System lab, University of Waterloo, Canada, https://uwaterloo.ca/mechatronic-vehicle-systems-lab/), there is a predesigned place for connecting a "Transmission Aperture" PTO to drive auxiliary devices.

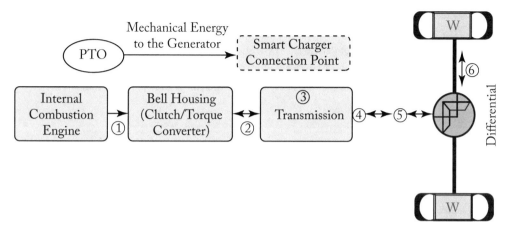

Figure 2.3: Schematic of the PTO configuration.

Figure 2.4: Installed "Transmission Aperture" PTO.

2.1 MODELING

Linking the model of powertrain components (e.g., differential, transmission, PTO), electrical parts (generator, batteries), and the engine will create whole system model. The model needs to be scalable so that the optimizer can change component sizes and find the optimal solution. Modeling approaches that could be used are [27] as follows.

(a) Forward-looking: As illustrated in Figure 2.5a, modeling and simulation start from the driver's point of view. The movement demanded power is sent to the powertrain parts model, which determines the wheels speed.

(b) Backward-looking: Known drive cycle data will create the required power cycle. From Figure 2.5b, energy demand is determined and transferred from the wheels to the main power source (engine) utilizing the powertrain components model. Considering the components' efficiencies, the power needed for each component is determined. Detailed dynamics of components is not considered; however, less complicated models are created.

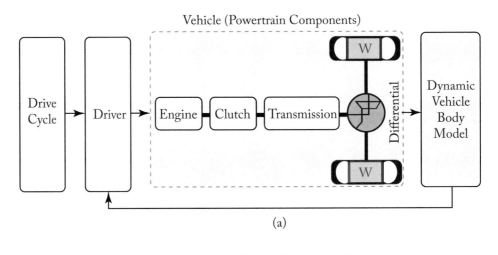

(a)

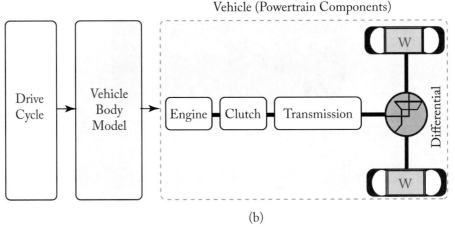

(b)

Figure 2.5: (a) Forward-looking vehicle model; (b) backward-looking vehicle model.

This study uses the modeling approach proposed by Guzzella and Rizzoni [28], which is a scalable backward-looking method to model all the related components. For each component, the demand power is determined by multiplying component current velocity and torque considering the component's efficiencies. Only the models of the main components are presented; for more detailed information, please refer to [9].

2.1.1   VEHICLE LONGITUDINAL DYNAMICS

A simple longitudinal vehicle model will fulfill the need of the backward-looking approach. Demanded longitudinal velocity (V_{Des}) and acceleration (A_{Des}) of the vehicle are fed to the axel-wheel model and the output will be the angular velocity (ω_{Wheel}), and the torque (T_{Wheel}) at the drive wheels [29]:

$$\omega_{Wheel} = \frac{V_{Des}}{R_{Eff}} \tag{2.1}$$

$$T_{Wheel} = F_{Total} R_{Eff}, \tag{2.2}$$

where

$$F_{Total} = F_{Drive} + F_{Drag} + F_{RR} \tag{2.3}$$

and

$$F_{Drive} = M_{Total} A_{Des} \tag{2.4}$$

$$F_{Drag} = \frac{1}{2} C_D \rho (V_{Des})^2 A \tag{2.5}$$

$$F_{RR} = C_{RR} M_{Total} g \cos \alpha \tag{2.6}$$

in the above equations, F_{Total}, F_{Drive}, F_{Drag}, and F_{RR} represent the total longitudinal force, driving force, aerodynamic drag force, and the rolling resistance, respectively. C_D and C_{RR} denote the drag and the rolling resistance coefficient. ρ and g indicate the air density and gravitational acceleration. A refers to the frontal area, and α means the road grade angle. R_{Eff} denotes the effective tire radius and M_{Total} indicates the total vehicle mass:

$$M_{Total} = M_{Veh} + M_{EES} + M_{Cargo}, \tag{2.7}$$

where M_{Veh} denotes the mass without EES packs and cargo. M_{Cargo} means the cargo weights and M_{EES} refers to the mass of the smart charger added (mainly the EES and its accessories). The total demand drive power (P_{Drive}), the power required to move the vehicle, can be determined from the following equation, where P_{Lost} represents the lost power of the powertrains:

$$P_{Drive} = \omega_{Wheel} T_{Wheel} + P_{Lost}. \tag{2.8}$$

2.1.2   GENERATOR MODELING

Generator electrical power (P_G) is in the form of the output voltage (V_G) and output current (I_G) as [30, 31]:

$$P_G = I_G V_G = \eta_G T_G \omega_G. \tag{2.9}$$

In which T_G and ω_G denote the input torque and angular velocity to the generator, and η_G shows the generator efficiency. During braking, regenerative braking power is transferred to the serpentine belt. However, due to safety, generator size restriction, and parts' efficiency, only a portion of the braking energy reaches the generator. The regenerative braking ratio factor (α_{Reg}) shows this portion. During braking, in the serpentine belt configuration:

$$T_G = \frac{\alpha_{Reg} T_T}{N_G} \tag{2.10}$$

$$\omega_G = N_G \omega_T \tag{2.11}$$

in which N_G denotes the generator ratio because of the serpentine belt pulleys. In the PTO configuration:

$$T_G = \left(\frac{\alpha_{Reg} T_T}{N_{PTO}}\right) \eta_{PTO} C_{ACT-PTO} = T_{PTO-out} \tag{2.12}$$

$$\omega_G = N_{PTO} \omega_T = \omega_{PTO-out}. \tag{2.13}$$

When the vehicle moves or even stops, the engine should charge the battery if its capacity drops to its minimum limit. In these scenarios, total demand engine torque (T_E) will be the addition of engine direct charging demanded torque (T_{E_Charge}), demand transmission torque (T_T) and T_R, indicating the required torque to overcome the engine resistance. Therefore:

$$T_E = T_R + T_T + T_{E\_Charge} \tag{2.14}$$

$$\omega_E = \omega_T. \tag{2.15}$$

In which ω_E is the engine's angular velocity. In no braking situation, in the serpentine belt configuration:

$$T_G = \frac{T_{E\_Charge}}{N_G} \tag{2.16}$$

$$\omega_G = N_G \omega_E \tag{2.17}$$

and in the PTO configuration:

$$T_G = \left(\frac{T_{E\_Charge}}{N_{PTO}}\right) \eta_T \eta_{PTO} C_{Act-PTO} \tag{2.18}$$

$$\omega_G = N_{PTO} \omega_E. \tag{2.19}$$

2.1.3   BATTERY MODELING

Batteries store the produced energy during regenerative braking or direct charging from the engine and power the auxiliary devices during the period it is supposed to be idling. A detailed battery model is complex and not necessary. The characteristics of the battery performance can be acceptable for modeling in HEVs. Changes in the level of energy and power of battery will be the main concern. Therefore, electric-based modeling method is selected for this study. Among electric-based models, "Dual Polarization (DP)" and "Thevenin" models have the best performance, however, both of them are created utilizing internal resistance model or "Rint." Dynamic voltage performance of the battery is ignored in the Rint model, however, it performs closely to the "Dual Polarization (DP)" and "Thevenin" models [32]. Moreover, the Rint model will not increase the calculations process and does not need an identification process to identify the battery model parameters such as resistor and capacitor. In the "Rint" model, current is determined based on charge-discharge power and internal resistance. Desired changes of the battery power ($P_{B\_Desired}$) are calculated by:

$$P_{B\_Desired} = R_{In}I_B^2 + V_{B\_OC}I_B, \tag{2.20}$$

where I_B, R_{In}, and V_{B_OC} indicate the current, internal resistance, and open circuit voltage, respectively. I_B and $P_{B\_Desired}$ will be positive during battery charging and will be negative during discharge cycles. By solving the above equation, I_B can be obtained:

$$I_B = \frac{-V_{B\_OC} \pm \sqrt{V_{B\_OC}^2 + 4P_{B\_Desired}R_{In}}}{2R_{In}}. \tag{2.21}$$

Only the lower value is adopted since the larger value is higher than the ideal current (I_{B_Ideal}) in which there is no waste of energy due to internal resistance. Related changes of the battery SOC and open circuit voltage (V_{B_OC}) are linked through the battery look-up table. Actual battery power (P_{B_Actual}) is shown by:

$$P_{B\_Actual} = V_{B\_OC}I_B. \tag{2.22}$$

By integrating (P_{B_Actual}), changes in the battery energy SOC level can be determined.

2.1.4 ENGINE MODELING

Internal combustion engines take the fuel and burn it inside the cylinder. During the whole process the internal energy of the fuel is converted to the mechanical energy to drive the vehicle to move [33, 34]. Willan's line engine modeling method [35] is used thanks to its nature of scalability and composability. Its scalability feature makes it independent of component size and the created model is able to be efficiently updated for other engines. The composability feature makes it possible that the components can simply link to the other related parts. In a general Internal Combustion Engine the following equation is valid:

$$\omega_E T_E = \eta_E P_{Fuel} = \eta_E \dot{m}_F H_L, \tag{2.23}$$

where P_{Fuel} is the enthalpy flow with the fuel mass flow $\dot{m}_F$. η_E refers to the engine efficiency, and H_L denotes the fuel's lower heating value. According to (i) mean effective pressure (p_{ME}

engine's ability to generate mechanical work) and (ii) fuel available mean effective pressure (p_{MF}, the maximum mean effective pressure generated by an engine with 100% efficiency using a unit fuel) engine efficiency is determined. In a steady-state active engine:

$$p_{ME} = \frac{N\pi}{V_{ED}} T_E \tag{2.24}$$

$$p_{MF} = \frac{N\pi H_L}{V_{ED}} \frac{\dot{m}_F}{\omega_E}. \tag{2.25}$$

N refers to the stroke numbers and V_{ED} is the engine's volume. According to the theory of thermodynamic efficiency and internal losses:

$$p_{ME} = e_E p_{MF} - p_L, \tag{2.26}$$

where e_E denotes the thermodynamic properties of the engine related to the pressure. p_L indicates the engine loss and is expressed by:

$$p_L = p_{LG} + p_{LF}. \tag{2.27}$$

In the above equation, p_{LG} and p_{LF} denotes the engine loss caused by the gas exchange and friction, respectively:

$$p_{LF} = k_1 \left(k_2 + k_3 \left(S\omega_E \right)^2 \right) \prod_{\max} \sqrt{\frac{k_4}{B}}, \tag{2.28}$$

in which B means the engine cylinder bore and $\prod_{\max}$ denotes the maximum boost pressure. S represents the engine stroke, and k are parameters to be determined experimentally. Considering the above equations, engine efficiency (η_E) and fuel mass flow ($\dot{m}_F$) can be defined as:

$$\eta_E = \frac{p_{ME}}{p_{MF}} \tag{2.29}$$

$$\dot{m}_F = \frac{\omega_E T_E}{H_L \eta_E}. \tag{2.30}$$

2.2   DESIGN OPTIMIZATION

No option is considered for the optimization of the existing vehicle drivetrain since it was considered that the proposed solution does not make critical changes in the existing vehicle. In the Multidisciplinary Design Optimization (MDO) approach, optimization methods (e.g., genetic algorithm (GA) [36, 37]) are used to handle complicated systems' design problems. These systems consist of different disciplines. Using MDO, it is possible to consider all disciplines simultaneously for better design efficiency. The smart charge system consists of different parts and disciplines; therefore, an MDO technique is essential for design optimization in this system to simultaneously optimize power management logic and component size. The optimal goal of the smart charger design is to: (i) maximize regenerative braking energy; (ii) better fuel efficiency; (iii) lower capital cost and changes in the original vehicle; and (iv) meet the auxiliary devices power demand at any given time. To achieve these, component (battery, generator) size and proper charging strategy variable (battery's SOC thresholds) are considered as the optimization parameters (design variable candidates).

2.2.1   OPTIMIZATION ALGORITHM STRUCTURE

The optimization process is performed by running these steps.

(a) "Initialization": Optimization algorithm updates the system models by transferring drive cycle, idling cycle, vehicle specification and design variable candidates (component sizes and SOC limits), as the input to the vehicle simulation model. Moreover, the lower and upper bounds of the design variable candidates, namely vector X_D, and the optimization method are defined in this step. As shown in Figure 2.6, at each iteration, the optimizer will choose a new X_D that contains optimization variables such as the powertrain component size and power management logic. This X_D should satisfy the optimization constraints, $C(X_D)$.

(b) "Vehicle Model Simulation": This model consists of various disciplines. The chosen X_D will be sent to this simulation model. Based on X_D, the initial conditions for the simulation will be updated. Utilizing the updated model as well as inputs, the vehicle operation is simulated to obtain the outputs of the system such as the fuel consumed by engine and electricity consumption through battery SOC.

(c) "Objective Function": The results of the simulation ($U(X_D)$) will be transferred to the objective function $J((X_D), U(X_D))$ to evaluate the objective function value (total system cost) for the given optimization variables.

(d) "Comparison": Optimizer compares objective function value J_{n+1} in the current iteration with the (J_n) of the previous iteration. Based on variations in the total cost (objective function value), the optimization algorithm chooses a new set of optimization variables (X_D) while taking the optimization constraints and terminating conditions into consideration. This process is repeated until the termination conditions are activated.

2.2.2   OBJECTIVE COST FUNCTION

The main challenge is to balance the cost and size of the battery and the remaining added components; with the costs of the consumed fuel and plug-in electricity consumed for driving vehicle auxiliary systems during idling. The goal of the optimization is to find the lowest total cost or highest saving by utilizing the smart charge system. The cost function should consider the capital and running costs of the smart charger over a given period. The objective function of the

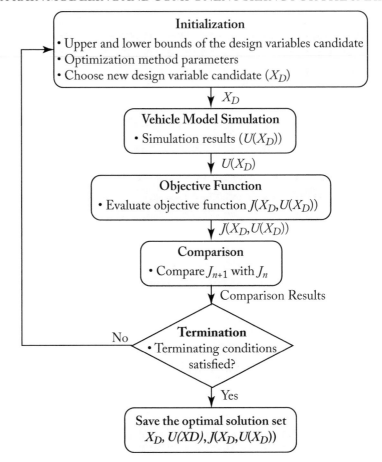

Figure 2.6: Optimization process.

optimizer is defined from a financial cost perspective to minimize: (i) the initial cost of batteries, generator, and other added accessories to the existing vehicle; (ii) the cost of fuel consumption for running auxiliary devices (like AC or heater); and (iii) the plug-in electricity consumption.

Based on the lifespan of battery packs in HEVs and EVs, it is assumed that a battery pack can generally last about five years [38]. In the optimization process, the optimizer tries to determine the best battery pack size, best generator size, and the most appropriate power management logic (critical battery SOC thresholds) in order to minimize the objective function $J\left((X_D), U\left(X_D\right)\right)$, which is defined as

$$Minimize\ J\left(X_D, U\left(X_D\right)\right);\ w.r.t.\ X_D \tag{2.31}$$
$$\text{subject to: } C\left(X_D\right)$$

in which optimization variables (X_D) are the number of battery packs, generator power, and battery SOC thresholds. The function $U(X_D)$ consists of vehicle model outputs. The constraints vector $C(X_D)$ represents the equality and inequality constraints of the problem. The objective function J is the total financial amount and energy consumptions, without considering the cost of the existing vehicle, and it is calculated over the expected lifespan of the smart charge system. Different parts of the objective function can be categorized as follows.

Fuel Cost

One of the benefits of using the Willan's line engine model is that this model can provide a good estimation of the consumed fuel based on the fuel energy density. Utilizing the above related equation, at each time step according to the value of the engine efficiency (η_E), the fuel heating value (H_L), the engine torque (T_E), and the engine angular velocity (ω_E), the value of the fuel mass flow ($\dot{m}_F$) can be calculated. Adding these values over time will result in the total fuel consumption of the vehicle over each driving cycle (active day). Total fuel consumption of the vehicle, $F(U(X_D))$, through the desired optimization process can be calculated as:

$$F(U(X_D)) = \text{Total Fuel Cost}$$

$$= \left[\sum_{t=0}^{T} Fuel_{Consumed} \times (Fuel_{Cost}) \right] \times Days_{Active} \times Years \qquad (2.32)$$

in which t is the vehicle model simulation time, T is the total time of simulation or drive cycle, and term $\sum_{t=0}^{T} Fuel_{Consumed}$ represents the total fuel consumption through one day (or one simulation). $Fuel_{Cost}$ is the unit price of fuel, $Days_{Active}$ is the number of active days per year, and $Years$ represents the targeted active years for the optimizer.

Plug-in Electricity Cost

It is considered that at the end of the day the vehicle will return to the specific station, which has the accessories to recharge the batteries up to 95% of their capacities. The total consumed electrical energy during this time can be calculated based on the change in the SOC level of the battery pack during overnight plug-in charging (SOC_{charge}). Final SOC of all working days will be the same since the SOC starts from the same level of 95% each day and the same drive cycle is assumed for each working day. Total consumed electrical energy during each overnight plug-in can be determined as:

$$Plug\_In_{Energy} = 0.001 \times 0.01 \times \left(SOC_{charge} \times Capacity_{Total} \times V_{B\_Charge} \right), \qquad (2.33)$$

where $Plug\_In_{Energy}$ is the total kWh consumed electrical energy during each night, $Capacity_{Total}$ is the total capacity battery packs, and V_{B_Charge} is the charging voltage of the battery pack which is one of the battery's specifications. The value of $Capacity_{Total}$ is based on watt-hours [Wh], in order to find the value of $Plug\_In_{Energy}$ in kWh, which is the based price calculation unit, the coefficient 0.001 is added. Moreover, the coefficient 0.01 is added to change the value of SOC_{charge} from present to the acceptable ratio for energy consumption. The total plug-in electricity consumption by the smart charge system during its expected life ($Plug\_In_{Cost-Total}$) is defined as:

$$P\left(U\left(X_D\right)\right) = Plug\_In_{Cost-Total}$$
$$= \left[Plug\_In_{Energy} \times Electricity_{Cost}\right] \times Days_{Active} \times Years \qquad (2.34)$$

in which the $Electricity_{Cost}$ is the price of each kWh electricity, purchased, during the overnight plug-in charging. $Plug\_In_{Cost-Total}$ is a function of $U\left(X_D\right)$ and depends on simulation model outputs.

Battery Cost

Battery packs are usually the most expensive electrical component in HEVs and EVs. It should be mentioned that the Original Equipment Manufacturer (OEM) battery will not be utilized to power the auxiliary devices. The most popular batteries, for this study, are lithium-ion, lead-acid, and nickel-metal-hydride types. The first constraint in battery packs optimization is that the number of series cells in the battery pack is constant in order to hold the nominal voltage in the same range for all of the chosen battery packs. Therefore, the optimizer will determine the number of the parallel set of cells in the battery banks. Based on that the total cost of the battery pack can be calculated as:

$$B\left(X_D\right) = Total\ Battery\ Cost = Battery_{Banks} \times Battery_{Cost}, \qquad (2.35)$$

where $Battery_{Banks}$ is the number of the parallel set of battery banks (determined by the optimizer) and $Battery_{Cost}$ is the unit price of each battery bank.

Generator Cost

Based on the possible configuration for installing the generator (serpentine belt or PTO configuration), there are different solutions for generator optimization. If the serpentine belt configuration is considered, there is a critical space limitation. Therefore (based on the available space for generator installation and generator power demand), it is possible to: (i) replace the OEM generator with a stronger generator; (ii) use a dual kit generator connection to use two generators (add another generator to the OEM generator); or (iii) utilize a dual kit generator connection with two new-stronger generators and remove the OEM generator. If the PTO configuration is considered, there is more space available for connecting the generator to the PTO; however, there are still space and weight limitations. After selecting the configuration, the optimizer will select the generator size as the design variable candidates (X_D), considering space constraints. Based on this selection, the optimization algorithm will update the simulation model and objective function with the specification (cost, weight, and power) of the selected solution (generator).

Other Capital Costs

Other than the cost of fuel, plug-in electricity, battery packs, and a generator, there are other capital costs that should be considered. In many cases, due to lack of information, the value of these costs should be estimated based on the available online data and vehicle parts dealer websites. Maintenance cost, labor cost, initial parts (accessories, PTO, power electronics, etc.) cost, and possible modification to the vehicle cost are some examples. These costs vary from case to case based on the application. For instance, due to the sensitivity of lithium-based batteries to the temperature and overcharge conditions, utilizing these kinds of batteries results in the use of more expensive power management control and plug-in chargers. This extra cost will be compensated by their lower weight to capacity factor and their better performance.

Total Objective Function

The objective function J is defined as the total costs of the added components and energy consumptions without considering the cost of the original vehicle. This function is the combination of the fuel cost, plug-in electricity cost, the initial cost of battery packs, generator initial cost ($Generator_{Cost}$), and other capital costs ($Other_{Cost}$):

$$
\begin{aligned}
J = & \left[\left(\sum_{t=0}^{T} Fuel_{Consumed} \times Fuel_{Cost} \right) + \left(Plug\_In_{Energy} \times Electricity_{Cost} \right) \right] \\
& \times Days_{Active} \times Years + Battery_{Banks} \\
& \times Battery_{Cost} + Generator_{Cost} + Other_{Cost} + ERROR,
\end{aligned}
\tag{2.36}
$$

in which term *ERROR* is either a zero or a large number to control the solution of the optimizer. If *ERROR* is zero the solution is acceptable. Otherwise, e.g., when the demanded power is more than the maximum available engine power, *ERROR* takes a large number to prevent the optimizer from selecting that solution. All of the variables of the above equation are explained in Table 2.1. The objective function J can be rearranged as:

$$
\begin{aligned}
J\left(X_D, U\left(X_D\right)\right) = & a_1 F\left(U\left(X_D\right)\right) + a_2 P\left(U\left(X_D\right)\right) \\
& + a_3 B\left(X_D\right) + a_4 G\left(X_D\right) + b_4 + E,
\end{aligned}
\tag{2.37}
$$

in which, a_1, a_2, a_3, and a_4 represent the optimization weighting coefficients for the fuel consumption, plug-in electricity, battery pack, and generator, respectively. $G\left(X_D\right)$ represents the generator cost, b_4 models the other capital cost of the system, and E shows the error signal preventing the optimizer from selecting impossible solutions. Considering the fact that all the values in Equation (2.36) are normalized to the actual dollar values, the optimization weighting coefficients (a_1, a_2, a_3, and a_4) in Equation (2.37) are equal to one. It should be noted that these coefficients can be modified based on new consideration for the optimization such as air pollution or battery degradation.

2.2.3   OPTIMIZATION CONSTRAINTS

There are different limitations and constraints that should be considered for selecting the design variable candidates, namely vector X_D. The constraints vector $C(X_D)$ is the equality and inequality equations representing optimization constraints. These constraints can be because of mechanical and physical limitations, availability of materials, maintenance issues, cost considerations, or many other reasons. Each of the optimization valuables has its own constraints.

Charging Strategy Variable Constraints (Battery's SOC Thresholds)

Battery prices and lifespan are important in the total system cost. The depth of discharge (DOD), charge-discharge cycles, and temperature are the critical factors influencing the battery price and lifespan. Considering a lifespan of five years for the battery, especially a lithium-based one in HEV application, is an acceptable assumption. However, in order to have the expected lifespan, it is required to take the degradation impact on battery life cycle into consideration. As the first tool, the battery's DOD should be actively controlled by finding the optimal charging strategy variable (battery's critical SOC thresholds). Optimization boundaries for the related design variable candidate (battery's critical SOC thresholds or DOD thresholds) should be selected accurately in order to make sure that the battery cells do not need replacement during the expected working periods of the smart charger.

Table 2.1: Optimization total objective function description

Variable	Definition
t	Vehicle model simulation time
T	Total time of vehicle model simulation or drive cycle
$\sum_{t=0}^{T} Fuel_{Consumed}$	Total fuel consumption through one active day
$Fuel_{Cost}$	The unit price of the fuel
$Plug\_In_{Energy}$	Total kWh consumed plug-in electrical energy during each night
$Electricity_{Cost}$	Price of the purchased kWh electricity during night hours
$Days_{Active}$	The number of active days considered for the vehicle
$Years$	The targeted lifespan of the system for the optimizer
$Battery_{Banks}$	The number of the parallel set of battery banks (determined by the optimizer)
$Battery_{Cost}$	The unit price of each pack of battery bank or each series set of battery bank cells
$Accessory_{Cost}$	The total cost of added initial parts
$ERROR$	Zero or a large number

Battery Sizing Design Variable Constraints

As mentioned, the first constraint in battery packs sizing is that the number of series cells in the battery pack should be constant in order to hold the nominal voltage in the same range for all of the chosen battery packs.

The second constraint for the battery sizing comes from Equation (2.21), where the total value of the term under the radical should be always non-negative so that the current has a real value. The values of R_{In} and V_{B_OC} are always non-negative. However, during the discharge of the battery, the value of the battery power ($P_{B\_Desired}$) is negative. In this condition, $4R_{In}$ should be so low such that the term $V_{B\_OC}^2$ will always be larger than the absolute value of the term $4P_{B\_Desired}R_{In}$. The value of R_{In} will be low when there are enough parallel battery packs connected. Therefore, the second constraint is the minimum number of parallel battery cells. The *ERROR* term in Equation (2.36) will take a large number if in the discharge condition the value of $V_{B\_OC}^2$ becomes smaller than the absolute value of $4P_{B\_Desired}R_{In}$.

The last constraint for the battery sizing relates to the total output power of the battery pack. The battery pack should have enough parallel cells to provide the maximum demanded current (or power) of the idling devices at all times. The maximum continuous discharge current of the battery pack always should be greater than the maximum demanded idling current.

Generator Sizing Design Variable Constraints

As mentioned before, available space for installing the generator in different configurations is the main constraint for generator sizing. In addition, maximum weight and maximum current-power of the generator (due to safety or material's properties limitations) are the other constraints that should be considered for generator sizing.

2.3   DESIGN OPTIMIZATION

Anti-idling solutions are used in different cases. To show the impact of utilizing the optimal smart charge system for anti-idling, a refrigerated delivery truck service vehicle is selected as an example. The daily working cycle of this vehicle contains many loading/unloading stops, during which the engine is on to power auxiliary devices. Designing an anti-idling solution for such vehicles can provide large fuel savings.

"Ford Transit Connect Cargo XL-2010" is considered the targeted service vehicle. The specifications of this vehicle are presented in Table 2.2. A modified drive cycle, as demonstrated in Figure 2.7, is generated by replicating the "UDDS" typical driving cycle and adding idling durations in order to simulate the vehicle's full day active cycle. In this new cycle, the vehicle drives 258.564 km with a total idling time of 8,205 s. Based on the engine model of the vehicle, idling fuel consumption rate of the Ford Transit Connect Cargo XL-2010 can be estimated as 0.1432 [mL/s] and in this one-day drive cycle, 1.175 [L] gasoline will be consumed just for activating the engine during idling periods. If the auxiliary devices are activated during idling,

Table 2.2: Ford transit connect cargo XL-2010 specifications

Vehicle: Ford Transit Connect Cargo XL-2010	
Engine capacity [L]	2.0
Total vehicle mass without cargo [kg]	1,524
Cargo mass [kg]	500
Tire efficient radius [cm]	30.76
Tire mass [kg]	20
Air density [kg/m^3]	1.2
Vehicle frontal area [m^2]	3.6171
The drag coefficient of the vehicle	0.32
Rolling resistance coefficient	0.01

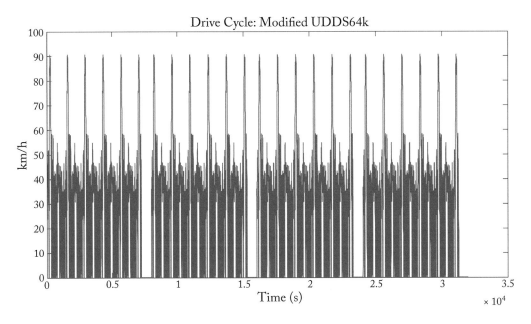

Figure 2.7: Modified drive cycle for simulation of a whole day active cycle of the vehicle.

fuel consumption rate increases to 0.3483 [mL/s] and in this one-day drive cycle, 2.585 [L] gasoline will be consumed.

A simulation model of a conventional vehicle (vehicle model without auxiliary power, extra battery packs, extra cargo, and regenerative braking system) over one-day drive cycle, results in 37.690 [L] gasoline consumption per day. Considering 1.175 [L] of this amount is for idling condition, 40.281 [L] gasoline is used for vehicle displacement (drive cycle) per day. If an on-off auxiliary power of 3 kW, as shown in Figure 2.8, is added to the previous conventional vehicle model, a service vehicle without an anti-idling solution is modeled. In this vehicle, when the auxiliary power is on, the amount of demanded auxiliary torque in each engine speed is calculated and added to the regular torque of the conventional vehicle during driving or idling. The impact of auxiliary power (service cycle) fuel consumption is shown in Table 2.3. Auxilary power consumes about 15% of the total fuel.

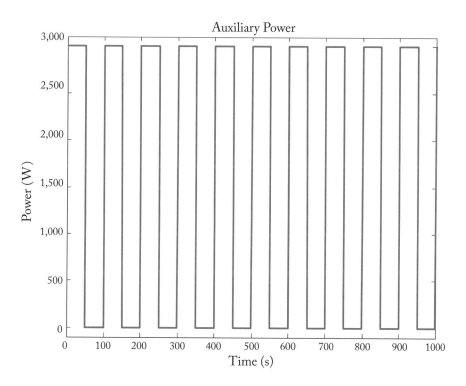

Figure 2.8: Demanded auxiliary power.

If the model of battery packs and regenerative braking system is added to the previous model (conventional service vehicle without anti-idling) a service vehicle with an optimal smart charger anti-idling solution is modeled. In this model, demanded auxiliary power is provided by the stored energy of the added battery packs which are charged using regenerative braking energy (wasted energy in conventional vehicles), plug-in electricity energy (cheaper than the fuel used in the engine to provide the same amount of energy), and directly from the engine, but mostly in high engine efficiency conditions. There is a balance between the number of the battery packs and the total system cost.

Table 2.3: Conventional vehicle simulation results

Conventional Vehicle	Unloaded Vehicle			Loaded Vehicle (500 kg Cargo)				
	Idling	Drive Cycle	Total Without Auxiliary	Idling	Drive Cycle	Total Without Auxiliary	Service Cycle	Total With Auxiliary
Fuel Consumption [L]: Gasoline Per Day	1.175	36.515	37.69	1.175	40.281	41.456	7.402	48.857
Fuel Cost: [$] for 5 Years	2,290.3	71,176	73,466.3	2,290.3	78,548.7	80,839	14,433.1	95,271.4

Different optimization methods can be used to optimize the number of battery packs and the critical battery *SOC* threshold. In this study, different models of lithium-ion and lead-acid (dry cell) batteries are chosen for the component-sizing problem. The specifications of the chosen batteries are presented in Table 2.4. The results of these optimization procedures for the six chosen battery models are also shown in Table 2.5. Detailed financial analysis of smart charge system equipped vehicle is presented in Table 2.6.

Table 2.4: Chosen battery packs specifications

Battery Type	Nominal Voltage [V]	Capacity [kWh]	Capacity [Ah]	Weight [kg]	Price [$/Unit]
EV12_140X DiscoverDryCell	12	1.68	140	50	490
EV12_180X DiscoverDryCell	12	2.172	181	60	585
EV12_8DA_A DiscoverDryCell	12	3.2	260	78	740
EV12_Li_A123 ALM12V7	13.2	0.06	4.6	0.85	125
EV12_Li_A123 ANR26650	13.2	0.029	2.3	0.35	18
EV12_Li GBS_100Ah	12.8	1.28	100	12.8	555

Table 2.5: Optimization results for six different battery models

Battery Model	EV12_140X DiscoverDryCell		EV12_180X DiscoverDryCell		EV12_8DA_A DiscoverDryCell		EV12_Li A123_ALM12V7		EV12_Li A123_ANR26650		EV12_Li GBS_100Ah	
Optimization Method	GA	SA	GA	SA	GA	SA	GA	SA	GA	SA	GA	SA
Optimization Time [s]	20,570	43,685	21,279	43,732	22,422	43,036	22,258	43,591	21,099	44,180	20,943	44,249
Critical Level Battery SOC %	70.83	70	72.5	70.6	70.45	70	30.74	30.16	33.55	30	31.34	33.16
Number of Battery Packs	3	3	2	2	2	2	94	84	37	37	1	1
Total System Cost ($)	92,165	92,175	92,147	92,073	92,100	92,107	107,775	107,960	92,593	92,418	92,982	93,123

DOD feature of lithium-based batteries and their low weight compared to the same capacity lead-acid batteries makes the lithium-based batteries an interesting choice. However, for this example the number of battery packs are low; therefore, the higher initial cost of lithium-based batteries (especially "EV12_Li_A123 ALM12V7") increases the total system cost and the benefits of lithium-based batteries cannot compensate this extra costs. This is the reason for the negative value in the "Total Saving by Utilizing the smart charging system [$]" row in Table 2.6.

2.4   SUMMARY

To maximize the benefits of the smart charge system, it should be optimized at the system level. Since there are different parts with different disciplines in the smart charger, a Multidisciplinary Design Optimization (MDO) technique should be utilized. The first step is to develop a scalable and composable system model that can be used by the optimizer algorithm. The MDO optimizer simultaneously optimizes the component sizing and power management logic in the overall system. The optimization objective function is defined from a cost perspective, where the objective is to minimize the total cost of the system through balancing the size of the batteries-generator and proper charging strategy while minimizing the fuel consumption, overnight plug-in electricity, and capital cost of other accessories. This process is implemented for designing an anti-idling system for a "Ford Transit Connect" service vehicle. It is shown that

Table 2.6: Optimization results financial analysis

Battery Model		EV12_140X Discover DryCell		EV12_180X Discover DryCell		EV12_8DA_A Discover DryCell		EV12_Li_A123 ALM12V7		EV12_Li_A123 ANR26650		EV12_Li GBS_100Ah	
Optimization Methods		GA	SA	GA	SA	GA	SA	GA	SA	GA	SA	GA	SA
Added System Cost	Number of battery packs	3	3	2	2	2	2	94	84	37	37	1	1
	Unit battery pack price [$]	490	490	585	585	740	740	125	125	18	18	555	555
	Total ESS price [$]	1,661	1,661	1,322	1,322	1,672	1,672	13,278	11,865	3,010	3,010	2,509	2,509
	Total initial parts cost [$]	2,000	2,000	2,000	2,000	2,000	2,000	3,600	3,600	3,600	3,600	3,600	3,600
	Total initialization cost [$]	3,661	3,661	3,322	3,322	3,672	3,672	16,878	15,465	6,610	6,610	6,109	6,109
	Total plug-in electricity cost [$]	163.9	197.7	173.7	216.5	233.6	274.8	656.1	596.3	418	449.6	556	537
Battery packs, plug-in electricity and initial system costs [$]		3,825	3,859	3,496	3,539	3,906	3,947	17,534	16,061	7,028	7,060	6,665	6,646
Fuel Saving [$]	Conventional vehicle fuel cost [$]	95,271	95,271	95,271	95,271	95,271	95,271	95,271	95,271	95,271	95,271	95,271	95,271
	Optimized vehicle fuel cost [$]	88,340	88,316	88,651	88,535	88,194	88,160	90,241	91,899	85,564	85,358	86,317	86,478
	Total fuel saving in five years [$]	6,931	6,955	6,620	6,736	7,077	7,111	5,030	3,372	9,707	9,913	8,954	8,793
Total Saving by utilizing the smart charger [$]		3,106	3,096	3,124	3,197	3,171	3,164	-12,504	-12,689	2,679	2,853	2,289	2,147
Investment returning time [Year]		2.641	2.632	2.509	2.466	2.594	2.582	16.78	22.93	3.405	3.334	3.411	3.474

the optimal smart charger meets the auxiliary power demand in all time, eliminates the vehicle idling, and has a return on investment timeframe of 2.5–3 years.

CHAPTER 3

Driving and Service Cycle Estimation

In this chapter, as the prerequisites, it is beneficial to predict the driving and service cycles, used for the power management strategy development. Driving cycle and service (duty) cycle, in this study, refers to the velocity and auxiliary load information with respect to time history.

3.1 PREDICTION OF THE AUXILIARY LOAD IN SERVICE VEHICLES

As mentioned previously, the power management controller of the smart charger systems can benefit from *a priori* knowledge on the auxiliary load (duty cycle) for optimal performance. This section reviews a method proposed in [39] that predicts the distance of the places where a vehicle stops for a long time and the corresponding demand auxiliary power during these stops.

A service vehicle usually stops at the same locations daily; therefore, it is much easier to predict future information through historical data when compared to other vehicles. One option to predict these parameters is to use the duty cycle on the last working day. However, for the sake of a more realistic forecast, data collected over the past four weeks is suggested to be used. Furthermore, a greater weighting factor may be applied to recent data points to account for new changes to route and demand auxiliary load of the service vehicle:

$$\delta_{pred} = \frac{a_1\delta_1 + a_2 \sum_{i=2}^{8} \delta_i + a_3 \sum_{i=9}^{29} \delta_i}{a_1 + 7a_2 + 21a_3},$$

(3.1)

where δ_{pred} means the estimated parameter for the next working day. δ_i denotes recorded parameter in the i_th day ($i = 1$ and $i = 29$ indicate the first and last day of the data collecting period, respectively), and $a_1 > a_2 > a_3$ are weighting factors, which may be tuned heuristically for each specific vehicle. For example, if the route and stop durations of a vehicle rarely changes, the weights are chosen similarly, indicating a variation in recently collected data should not noticeably affect the forecast. However, if the stopping time and route vary regularly, the weights on the recently collected data points must be larger than the previous ones, to renew predictions more promptly.

The average demand auxiliary power at a stop location is forecasted by Equation (3.1), which could be also adopted to predict the stop duration and location. However, this method cannot determine whether a new change in the locations and durations of stops will be repeated in the future or not. As a result, an intelligent approach is necessary to predict if the change in the duty cycle of the vehicle is permanent or temporary. To realize this purpose, a clustering method is proposed, which sorts the collected data points for each stopping location into a group. Once the number of data points collected in the same stop location is greater than a predefined value, a new cluster should be added by the algorithm, which represents a new stop location. On the other hand, when the data points in a cluster drop below the predefined threshold, the algorithm should eliminate the corresponding cluster to illustrate the vehicle would not stop at that old location.

Two most widely used clustering techniques are k-mean [40, 41] and Density-Based Spatial Clustering of Application with Noise (DBSCAN) [42]. The k-mean clustering is a centroid model that seeks to divide n data points into k groups such that each data point is part of the cluster with the nearest center (mean). The k-mean algorithm operates by taking following procedure.

1. Choose k data points as the center of clusters randomly.

2. Find the Euclidean distance between all cluster centers and each data point.

3. Assign each data point to a cluster that has a minimum distance to its center.

4. Calculate new cluster centers by finding the mean value of all the data points in each cluster.

5. Calculate the Euclidean distance between new cluster centers and each data point.

6. If any data point is assigned to a new cluster center, repeat from step 3; otherwise, stop.

The main drawback of the k-mean technique for the problem on hand is that it needs a predefined number of clusters. This means that the number of engine-OFF stops should be known in advance, whereas the goal is to add and remove stop locations automatically. The other disadvantage of this technique is the inability to handle noisy data or outliers. Consequently, if the vehicle stops in a location only one time, this data point affects the clustering process.

To overcome the problems associated with the k-mean technique, the DBSCAN method, which is a density-based model, is utilized. This technique does not require a predetermined number of clusters, and it detects outliers (noise points) efficiently. In this method, data points are categorized into three groups: core points, border points, and noise points. A core point, as shown in Figure 3.1, is a point where more than a minimum number of data points (*MinPts*) is available in its *Eps*-neighborhood. A border point has fewer points than *MinPts* in its *Eps*-neighborhood, but it is in the neighborhood of a core point, whereas a noise point is neither a core point nor a border point.

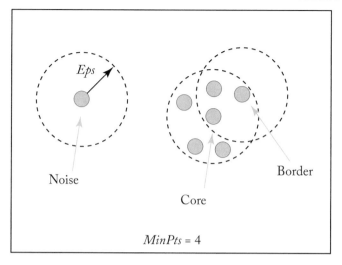

Figure 3.1: Definition of core, border, and noise points in DBSCAN method.

Definition: A point p is directly density-reachable from a point q if p is in the neighborhood of q and q is a core point. In addition, a point p is density-reachable from a point q if there is a chain of points $p_1, p_2, \cdots, p_n$ ($p_1 = p$, $p_n = q$) so that p_{i+1} is directly density-reachable from p_i (see Figure 3.2).

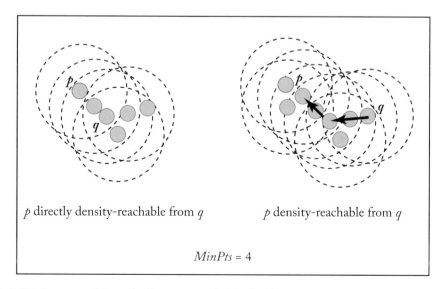

Figure 3.2: Definition of directly density-reachable (left) and density-reachable (right).

The algorithm of DBSCAN is presented in Figure 3.3a. Furthermore, a simple instance of this clustering method is indicated in Figure 3.3b. Interested readers are referred to [42] for a detailed description of this method.

```
for each data point p
    if p is not yet in a cluster then
        if p is a core point then
        all data point that are density-reachable from p are assigned to a new cluster
        else
        assign p to Noise point
        end
    end
end
```

(a)

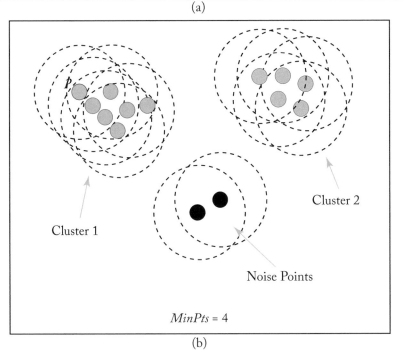

(b)

Figure 3.3: (a) Algorithm of DBSCAN and (b) an example of DBSCAN clustering.

Eps and *MinPts* are two parameters that should be determined by a heuristic approach for the DBSCAN. The *Eps* value needs to be tuned by taking two aspects into account. First, it has to be reasonably selected to recognize noise points. If the *Eps* is very large, then the noise points will be assigned to the clusters, leading inaccurate prediction. Second, the *Eps* should be determined so that the cluster number matches the number of engine-OFF locations. That means if the *Eps* is very small, the corresponding data points of an engine-OFF location may be divided into multiple clusters. On the contrary, a large *Eps* may put the data points of multiple engine-OFF locations in only one cluster. The *MinPts* value mainly affects update rate of stop locations. If a small value is selected for the *MinPts*, a new cluster is created only a few days after the vehicle stops in a new location. A big value for the *MinPts*, however, removes old stop locations slowly. Therefore, it takes a while to update stop locations when a vehicle does not stop at the old location. It should be noted that creating new stop locations has a higher priority than removing old ones because the battery might lack energy for auxiliary devices when a stop location is not predicted. As a result, it is recommended to choose a small value for the *MinPts* to ensure that all possible stop locations are predicted.

By implementing the DBSCAN on the historically collected data points, durations (as well as locations of the stops) are forecasted by determining the mean of the points in the clusters. Data points are multiplied by the weighting factors indicated in Equation (3.1) to put more value on recently recorded data. For instance, the collected data points (i.e., locations and durations of the stops) of a truck are illustrated in Figure 3.4. The weight factors, *MinPts* and *Eps*, are chosen as in Table 3.1.

Table 3.1: Parameters in DBSCAN method

Symbol	Parameter	Value
α_1	Weight for the data collected in the previous day	5
α_2	Weight for the data collected in the fourth week	3
α_3	Weight for the data collected in the first three weeks	1
MinPts	The minimum number of data points in a cluster	10
Eps	Neighborhood radius	5

As presented in Figure 3.4, the daily stop locations of the truck are almost similar. The recorded data points are sorted into seven groups, shown in solid circles, by the DBSCAN approach. The result concludes that besides seven permanent stop places, an additional stop is expected in the next working day (indicated by a new cluster in the dashed circle). It also shows there is a low possibility of stopping at an old location (indicated by the dotted circle). To determine the duration and location of each stop, center of the corresponding cluster needs to be calculated. A cluster center is the mean of all its data points, and the points that are not part of any cluster are noise points, which have no effect on other clusters.

Considering a service vehicle has a similar duty cycle every day, this prediction method is reliable. Although the prediction can be updated by any available real-time information to improve the performance of the power management controller. For example, in refrigerated delivery trucks, the variation of freight mass, ambient temperature, solar radiation, etc., can change the duty cycle considerably. Therefore, the prediction of the duty cycle can be updated

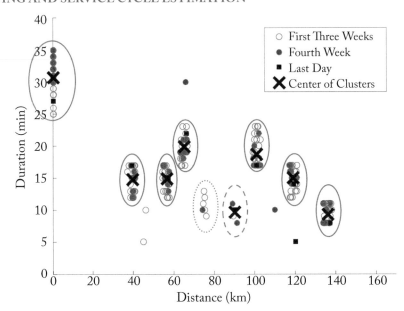

Figure 3.4: Forecast of engine-OFF locations and durations based on the DBSCAN approach.

online by a mathematical model that takes all these factors into account. The prediction of stop locations can be also updated by driver's inputs and a navigation system.

3.2   DRIVING INFORMATION ESTIMATION

In the previous section, the stops and the auxiliary loads are identified from the historically collected data. In this section, the method proposed in [43] is reviewed, where the driving information, i.e., vehicle velocity, in the near future is predicted by using the historical data as well. Many different methods used for predicting the future driving information has been used in current literature, and the most popular ones are listed in Table 3.2. For the detailed information, please refer to [44]. The method used in this study is statistical.

Table 3.2: Prediction methods of the future driving information

Prescient	Frozen-time	Exponential-varying	Stochastic	Telematics	AI	Statistics
[45]	[46]	[47]	[48]	[49]	[50]	[51]

Use the UDDS, standing for urban dynamometer driving schedule, as an instance to represent the collected historical data. Figure 3.5 indicates the vehicle velocity, acceleration, and power during the first 20 min. The power in positive is used to move the vehicle, while the negative one indicates the power needed to brake the vehicle, which can be potentially recaptured during braking.

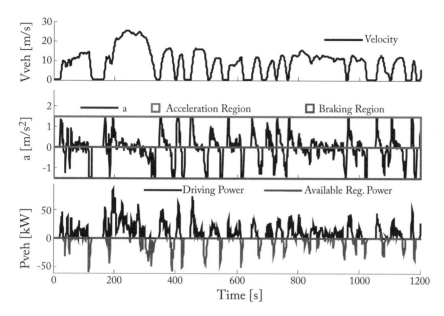

Figure 3.5: Driving information of the first 20 min in UDDS.

From Figure 3.5, it is concluded that both the negative and positive power appears alternately; in the meantime, the amplitudes are relatively identical in most regions. Ideally, it is feasible to adopt a window with a proper width to move along the data, and the averaged power in the covered area can be obtained. In other words, since the driving power and the regenerated power is going to and from the energy storage (e.g., a battery to maintain its state) this average method can be considered as both powers counteract with each other in the power level instead of in the SOC level.

In Figure 3.6, the solid blue window denotes the current location of the moving window, and the dotted one shows the previous time instant, where w and δt refer to the width and moving stride, respectively. Once the window shifts each time instant, the averaged vehicle power in the window is used as the vehicle power at the current point.

In this application, the moving stride δt is fixed at 1 s and the window width is tested by three different values (i.e., 10, 50, and 100 s). The resulting vehicle power information is demonstrated in the following plots, where both the averaged regenerative and driving power are calculated by moving the window.

Figures 3.7, 3.8, and 3.9 show that as the width of the moving window increases, the average power used to drive and brake the vehicle tends to be smoother with small fluctuations. Then, a window with 100 s width is selected and studied in the controller design process by

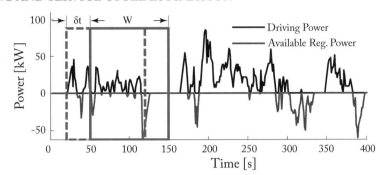

Figure 3.6: Driving and braking power with the moving window.

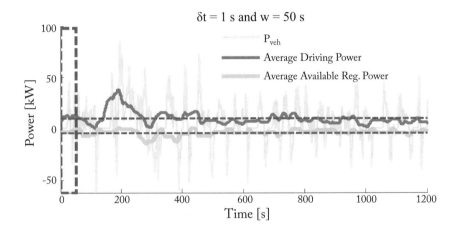

Figure 3.7: Driving and braking with a 50 s window.

balancing the computational load and prediction accuracy. As a prerequisite when designing the MPC power management controller, the vehicle power in the near future or the prediction horizon must be identified beforehand. In this study, inspired by the frozen-time MPC [45], such information in the whole prediction horizon is obtained through reflecting the fresh historical data. More specifically, the driving and braking power in the future 100 s will be mirrored by the one in the latest 100 s indicated in Figure 3.10.

3.3   SUMMARY

In this chapter, the auxiliary loads and truck stops are identified first by the DBSCAN based on the historical data. In addition, the driving information is predicted by the statistics-based method (i.e., averaging method) to analyze fresh historical data as well. Such information is

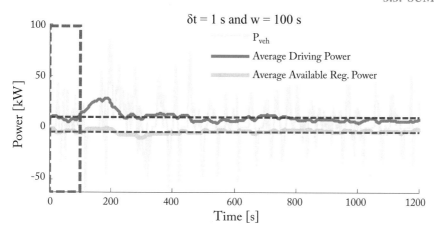

Figure 3.8: Driving and braking with a 100 s window.

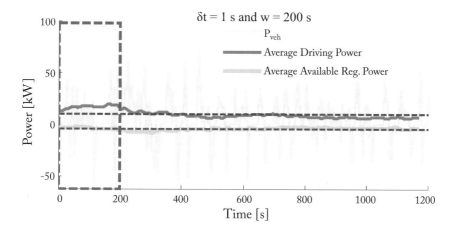

Figure 3.9: Driving and braking with a 200 s window.

necessary for the advanced PMS development, the effectiveness of these prediction methods together with the PMSs are verified in the next chapter.

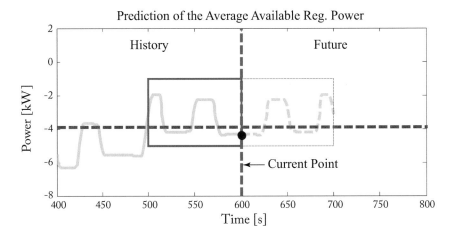

Figure 3.10: Vehicle power estimation.

CHAPTER 4

Power Management Controller Design for the Smart Charger

As a hybrid system, the updated powertrain with added smart charger is a complicated electro-mechanical-chemical system. The power flows between the components, the potential improvement of the fuel economy, and emission reduction for hybrid powertrains depends on the power distribution within the hybrid powertrains. Strategies that determine such power distribution are usually referred to PMSs. In this chapter, several different PMSs are studied to verify their feasibility and potentials to be used in the developed smart charger.

4.1 TWO-LEVEL DP-ADAPTIVE ECMS POWER MANAGEMENT CONTROLLER

Specifically, the PMS of the smart charger should satisfy the following requirements: to enable the regenerative braking feature, to ensure the batteries can supply electrical power to auxiliary devices at all locations where the engine is shut off, to decide the optimal ratio of the required power consumed by the auxiliary devices between the generator and battery to improve fuel economy, and to discharge batteries to the minimum allowed limit at the end of a trip. The last condition is applied only if plug-in capability, which allows the battery to be charged via the grid power, is available. Two options can meet all requirements, namely, the rule-based and the optimization-based approaches. For the former, it is not necessary to know the information beforehand (e.g., duty cycle of auxiliary devices, route, traffic information, duration and location of stops, etc.). As a result, to guarantee the battery has enough capacity before reaching a stop where the engine should be shut down, the control system charges batteries to the maximum allowed SOC once the engine is running and maintains the battery full until the next engine-OFF stop. The rule-based control strategies are easy to be applied but their performance may be far from optimal. Optimization-based control strategies, on the other hand, reach an optimal solution (charging/discharging strategy) to obtain better fuel economy. These control strategies are mainly divided into two classes: Global and Real-Time Optimization. The former needs full knowledge of future driving as well as highly computational efforts, but it can be used as the best benchmark to evaluate other counterparts. Dynamic Programming (DP) is one of the most popular techniques to obtain the global optimal solution for the power management controller. Real-time control optimization, however, uses past and current information without future information or possibly partial future information to minimize the cost function. ECMS is one of the widely used real-time schemes to determine the power flow in hybrid powertrain systems. The instantaneous summation of engine and equivalent battery fuel consumption is optimized [53]. An equivalent weight factor is required to calculate the equal fuel of the electrical power, which charges or discharges the battery. With a large selected value, the battery-discharging action is penalized and the engine is used more, whereas a small factor causes more electrical power to be drawn from the battery [54]. As a result, this parameter must be determined so that the battery is discharged to its minimum allowable limit when finishing a trip, which can be ensured by calculating the equivalent weight factor offline by an optimization method using *a priori* knowledge of the driving and duty cycle. To implement this strategy online, Adaptive ECMS (AECMS), which updates equivalent weight factors online, was presented [55].

To meet all the requirements of the smart charger, a novel two-level controller (DP-AECMS), which has been proposed by [56] is reviewed in this section. In the higher layer, a fast DP method is adopted to obtain optimal initial and terminal *SOC* for each segment based on available *a priori* knowledge. A segment, as illustrated in Figure 4.1, is defined as a duration of driving from an engine-OFF stop to the next one, or stopping time at a location for pick up/delivery where the engine is OFF. The main advantage of the higher level is that *SOC* trajectory can be adjusted online quickly when *a priori* knowledge such as the traffic condition is updated. In the lower layer, an AECMS regulates the split ratio of the demand auxiliary power between the battery and generator with the available information, such as the initial and terminal *SOC* of each segment calculated by the higher level of the controller.

4.1.1   HIGH-LEVEL CONTROLLER

In this layer, a fast DP method determines the amount of energy, which can be charged into or drawn from the battery during each segment. This controller also ensures the battery has sufficient energy at all possible engine-OFF stops. Since DP runs only at a few points (i.e., number of segments), the computational load is notably decreased, and an optimal solution can be obtained online momentarily.

In this DP technique, the optimal solution is determined for each segment rather than each time step. The system model is expressed by:

$$SOC(n + 1) = f\left(SOC(n), u_h(n)\right), \tag{4.1}$$

where $SOC(n)$ denotes the initial *SOC* of the battery at each segment. The model used to find the battery *SOC* in this level is defined as:

$$SOC(n) = \frac{E_{batt-rem}(n)}{E_{batt}}, \tag{4.2}$$

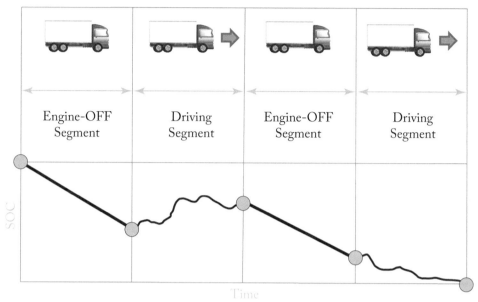

| Engine-OFF Segment | Driving Segment | Engine-OFF Segment | Driving Segment |

Initial and final SOC of each segment obtained by the high-layer control system

Real-time SOC obtained by the low-level control system

Figure 4.1: Definition of segments and SOC trajectory.

where E_{batt} means the total energy of the battery, and $E_{batt-rem}(n)$ is the energy left. In addition, $u_h(n)$ is the control input, determining the optimal split ratio:

$$E_{gen-seg}(n) = u_h(n)$$
$$E_{batt-seg}(n) = E_{aux-seg}(n) - E_{regen-seg}(n) - u_h(n), \qquad (4.3)$$

where $E_{gen-seg}$ is the total generated electrical energy via the generator in a segment utilized to power accessory devices and charge the battery. $E_{batt-seg}$ is the energy that is charged into or discharged from the battery in one segment, $E_{aux-seg}$ means the total demanded energy in one segment, and $E_{regen-seg}$ shows the total energy regenerated during braking in one segment. The goal is to obtain $u_h(n)$ so that the following cost function is minimized:

$$J = \sum_{n=0}^{N_s} , m_f(n), \qquad (4.4)$$

where (N_S) denotes the number of segments plus one and m_f is total fuel consumed by the engine in a segment. In order to obtain the amount of fuel consumed in each segment, the engine power at each time instant consumed to drive the generator and vehicle should be known. DP approach can find the amount of the generator energy but its distribution in the segment is not known, which is required to calculate the power of the generator. To overcome this issue, this energy is assumed to be equally distributed at each step in the segment. Even if this assumption may be invalid in determining the actual fuel consumption, it can still provide information accurate enough to assist in looking for a sub-optimal result in the high-level controller.

The fuel consumed at each segment must be evaluated many times to visit all potential results by DP algorithm, which can be very time-consuming. To address this problem, a histogram w.r.t. the engine torque and speed for each segment is created. This histogram, which can be obtained by prediction of the drive cycle, fit engine torque and speed into a 10-by-10 grid of equally spaced containers, as shown in Figure 4.2.

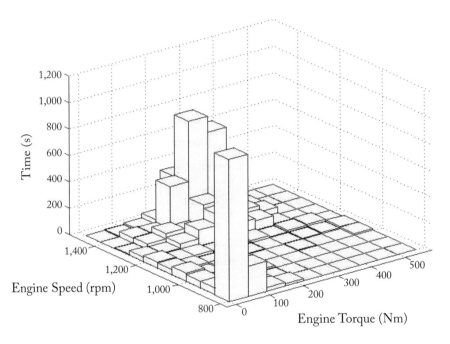

Figure 4.2: Histogram plot of engine torque and engine speed for a segment.

After that, the generator power is calculated by dividing the generator energy by the total time of the segment, and the generator torque is obtained by dividing the generator power by the speed index. In addition, by adding the generator torque to the torque index of the histogram, an updated torque index can be acquired, resulting in a histogram w.r.t. the engine speed and the new engine torque. Finally, the corresponding torque and speed of each bin are plugged into the engine fuel map to calculate the fuel rate. The fuel consumed by each bin is obtained by multiplying the fuel rate and the height, which denotes the period the engine works at a corresponding torque and speed. Then, the consumed fuel of each segment is obtained by summing the fuel used for all bins. Therefore, this approach decreases the computational efforts significantly, since only 100 points are utilized to determine the consumed fuel. A summary of the fuel consumption calculation is illustrated in Figure 4.3.

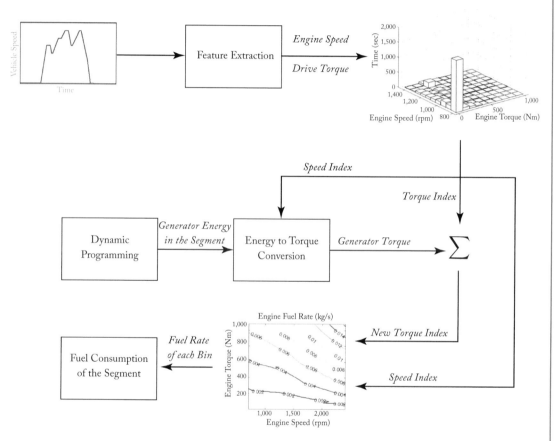

Figure 4.3: Fuel consumption for any segment in the higher-level controller.

Since the optimization is performed for each segment rather than at each time instant, the constraints on the generator torque and the engine torque are unable to be considered but will be applied to the lower layer. The only constraint that must be met in this layer is on the battery SOC:

$$SOC_{min} \leq SOC(n) \leq SOC_{max}, \tag{4.5}$$

where SOC_{min} and SOC_{max} are the lower and upper bound of the battery SOC. The Backward DP algorithm, which is described in [57], is used to determine the optimal control law $u_h^*(k)$ given the optimal state $SOC * (n)$.

4.1.2   LOWER-LEVEL CONTROLLER

In the previous section, the optimal initial and terminal SOC of each segment were obtained by the fast DP method. This guarantees that the battery is charged sufficiently to power auxiliary devices at stops and that it is fully discharged at the final state. Now, in the low-level controller, a real-time controller is needed to regulate the split ratio subjected to the desired initial and final SOC. To achieve this goal, the AECMS is adopted since it has shown promising performance in hybrid vehicles. The goal of the ECMS is to optimize the instantaneous equivalent fuel $(\dot{m}_{f,equ})$, which is expressed by [58]:

$$\dot{m}_{f,equ}(t, u_l(t)) = \dot{m}_f(t, u_l) + S \frac{P_{batt}(t, u_l)}{LHV}, \qquad (4.6)$$

where u_l denotes the ratio to be determined, $\dot{m}_f$ means the fuel rate of the engine, P_{batt} shows the power charged into or drawn from the battery. LHV refers to the lower heating value of the fuel, and S indicates the equivalent weight factor that converts the electrical energy in the battery to the equivalent fuel.

During each time step, all the potential $u_l(t)$ is inserted in Equation (4.6) to evaluate $\dot{m}_{f,equ}$ by the powertrain model. Then, $u_l^*(t)$, which associates with the minimum $\dot{m}_{f,equ}$, is chosen as the optimal result subjected to the following constraints:

$$\begin{aligned} SOC_{\min} &\leq SOC(t) \leq SOC_{\max} \\ T_{eng-\min} &\leq T_{eng}(t) \leq T_{eng-\max} \\ T_{gen-\min} &\leq T_{eng}(t) \leq T_{gen-\max}, \end{aligned} \qquad (4.7)$$

where $T_{eng-\min}$, $T_{eng-\max}$, $T_{gen-\min}$, and $T_{gen-\max}$ are the lower and upper bounds of the engine and generator torque, respectively.

By applying Pontryagin's minimum principle, it shows the optimal solution of the equivalent weight factor which can be obtained once the future information is known [58]. The optimal solution (s_{opt}) guarantees the terminal SOC reaches the desired value. However, an iterative optimization algorithm is used to obtain the equivalent weight factor, although it is not a computationally efficient approach, especially when the initial guess is far from the optimal solution. Consequently, it is not appropriate for real-time application since it takes a significant amount of time to converge once the future information is updated. AECMS is a method that can update the equivalent weight factor (s) in real time. Several different approaches for adapting the equivalent weight factor have been used, but the most common one uses the affine function of SOC error and a PI controller [59]:

$$s(SOC, t) = s_0 + k_p(SOC_{ref} - SOC(t)) + k_i \int_0^t (SOC_{ref} - SOC(t)), \qquad (4.8)$$

where s_0 denotes the initial guess of the equivalent weight factor. k_p and k_i are the proportional and integral gain, respectively. SOC_{ref} refers to the reference SOC, expressed by:

$$SOC_{ref} = SOC_0 - (SOC_0 - SOC_{ref}) \frac{D(t)}{D_{tot}}, \tag{4.9}$$

where SOC_0 and SOC_f denote the initial and terminal SOC of each segment, respectively. $D(t)$ indicates the traveled distance and D_{tot} means the total distance in a segment. The equation shows the SOC changes linearly w.r.t. the distance in a segment. As a result, it ensures the battery can be discharged or charged at the end of each segment as suggested by the high-layer controller.

4.1.3   CASE STUDY

A case is studied in this section to evaluate the advantage of implementing the DP-AECMS controller on a RAPS that is installed on a refrigerated delivery truck with the driving cycle and demand auxiliary power (duty cycle) shown in Figure 4.4. The truck has seven stops where the engine must idle to keep the refrigeration system running. A summary of the drive and duty cycles is available in Table 4.1.

Table 4.1: Summary of truck's drive and duty cycles

Parameter	Value
Total time	6.65 hrs
Total distance	171.7 km
Total idling time	2.25 hrs
Total auxiliary load	11.57 kWh
Vehicle mass	6,000 kg
Freight mass	1,500 kg

Using the MDO method, which was explained in the previous chapter, a 9 kWh lithium-ion battery and a 5.5 kW generator are used for the smart charger. Furthermore, the minimum and maximum allowed SOC of the battery are determined to be 30% and 80%, respectively.

Two scenarios are evaluated in this case study: the battery can be completely depleted when the trip is finished (plug-in capability is available), and the battery should be fully charged at the final destination (without plug-in capability). Figure 4.5 illustrates the battery SOC when DP-AECMS is utilized for the PMS. Simulation results demonstrate that RAPS can eliminate engine idling during stops by shutting down the engine and utilizing the battery to power the refrigeration system. Therefore, the emissions generated during engine idling are substantially reduced.

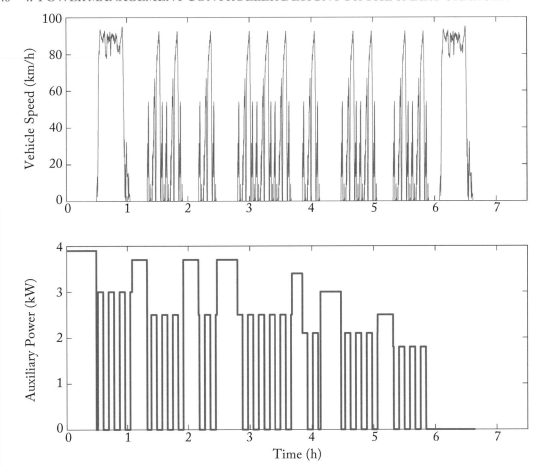

Figure 4.4: Vehicle speed (driving cycle) and demand auxiliary power (duty cycle) of the delivery truck.

As listed in Table 4.2, the fuel consumption of the conventional truck, where the compressor of the A/C-R unit is powered directly by the engine, is compared to that of the identical truck after the smart charger is added. In addition, this table presents a comparison of the performance of DP-AECMS, rule-based, and DP controller.

The smart charger with the DP-AECMS controller witnesses a 9.1% improvement in fuel consumption, in contrast to the conventional one (without the system). The enhancement is 5.1% with the assumption that the battery should be completely charged by the end of the trip (non-plugin). Furthermore, the rule-based controller showed only 2.3% enhancement in the fuel economy. The poor performance of the rule-based control strategy may be due to three possible reasons. First, the battery may be charged during the operation of the engine at low-

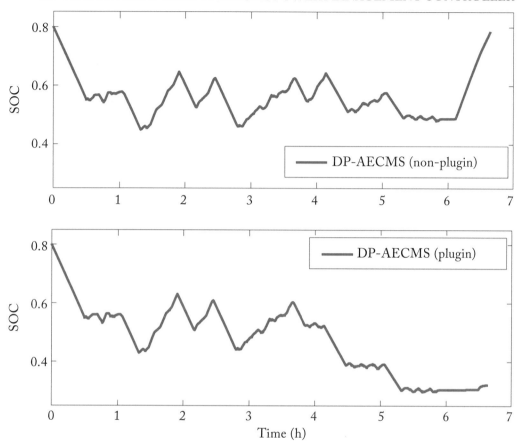

Figure 4.5: *SOC* of the battery when DP-AECMS controller (for both non-plugin and plugin) is utilized.

efficiency regions. Second, given that the battery *SOC* reaches its maximum allowable value, the regenerative braking is not working anymore when the battery is full. Last, the battery will not be discharged to the desired value at the final state where the truck has access to the grid power, resulting in more fuel consumption. The small difference between the truck's fuel economy from the DP-AECMS controller and DP (i.e., global optimal solution) verifies the good performance of the DP-AECMS controller in the proposed smart charger system.

Table 4.2: Consumed fuel and terminal battery SOC from different controllers in the smart charger

	Fuel Consumed	Fuel Saving	Terminal Battery SOC
Truck with no anti-idling system	32.04 L	-	-
Smart charger (Rule-based)	31.29 L	2.3%	80%
Smart charger, non-plugin (DP-AECMS)	30.34 L	5.1%	78.8%
Smart charger, plugin (DP-AECMS)	29.12 L	9.1%	32.2%
Smart charger, non-plugin (DP)	29.6 L	7.6%	80%
Smart charger, plugin (DP)	28.55 L	10.9%	31.6%

4.2   THE MPC-BASED POWER MANAGEMENT CONTROLLER

Recently, MPC has been extensively researched both theoretically and in real-world applications. The same thing also can be seen in the MPC-based PMSs development for hybrid powertrains. Hundreds of studies have been conducted and published from different perspectives. Reference [44] comprehensively reviewed the MPC-based PMSs in the current studies, where the MPC strategies can be categorized into the following several classes—methods to predict the future information, model types used in PMSs, and so on. For the definition and the applications of each method, please refer to [44] for more information. This chapter presents two types of the MPC-based PMSs: prescient MPC and the statistics-based MPC.

4.2.1   PRESCIENT MPC

Service vehicles, e.g., delivery trucks or public buses, usually drive on predefined routes; therefore, it is possible and beneficial to adopt the MPC algorithm to maximize the fuel economy of the proposed smart charger. Obviously, the mass/load of such vehicle is changing over a driving cycle and the fuel consumption calculation is heavily dependent on the vehicle total mass. Therefore, an algorithm to identify the vehicle total mass should be designed to deal with this variation. Even though the route is predefined, the vehicle may drive differently from it because of different traffic conditions in real-world situations. To address this issue, the MPC algorithm predicts its future behavior in a large step size to enhance its robustness. The designed MPC, which is adaptive to the mass/load variations, is compared to a prescient one under different cases to demonstrate its applicability and optimality. The developed approach does not rely on the powertrain configuration so it can be easily applied to other kinds of hybrid powertrains.

Vehicle Mass Identification

For the sake of completeness, in this section, a method for estimating the vehicle mass using a Kalman filter [61, 62] is discussed. Generally, a Kalman filter is used as a state observer. However, the parameter identification problem can be formulated in the state space form. To apply the Kalman filter, a parametric model should be firstly formulated by the component model presented in Chapter 2 in the form of:

$$
\begin{cases}
z = \theta(k)\phi(k) = (T_{eng} - T_{eng \to alt})\dfrac{N_{tf}\eta_{tran}}{R} - \dfrac{1}{2}\rho C_D A_f V_{veh}^2 \\
\theta(k) = M \\
\phi(k) = C_r g \cos\alpha + g \sin\alpha + a
\end{cases}
\tag{4.10}
$$

In the above equation, z denotes the process output, θ means the actual value of the parameter to be identified, and ϕ indicates the regression vector. Then, the information such as engine torque (T_{eng}), vehicle speed (V_{veh}), and engine speed (ω_{eng}) is acquired through the control area network (CAN). The alternator torque $(T_{eng \to alt})$ from the engine is obtained by the control input $(P_{eng \to alt})$. The acceleration information is collected by the accelerometer, which can also be obtained from the vehicle speed if an accelerometer is not available. By comparing the engine and vehicle speeds, the accumulated ratio of the final drive and transmission (N_{tf}) can be determined. Furthermore, the road grade (α) is assumed to be acquired by a GPS receiver. Other parameters, such as the efficiency of driveline (η_{tran}), the rolling resistance coefficient (C_r), the coefficient of aerodynamic resistance (C_D), the mass density of the air (ρ), and the vehicle frontal area (A_f) are supposed to be known and invariant w.r.t. time. Therefore, the parameter to be estimated is the vehicle mass (M). The aim is to find the best estimation of parameter to make the model output $(\hat{z})$ follow the process output (z) in best accuracy. In each time instant, the estimation process is renewed by

$$
\hat{\theta}(k) = \hat{\theta}(k-1) + K(k)e(k)
\tag{4.11}
$$

in which $\hat{\theta}$ represents the estimated values of the parameters, and e is the estimation error, reflecting the deviation between the process and the model output:

$$
e(k) = z(k) - \hat{z}(k);
\tag{4.12}
$$

moreover, K denotes the Kalman gain, which is determined by:

$$
\begin{cases}
K(k) = \dfrac{P(k-1)\phi(k)}{R_2 + \phi^T(k)P(k-1)\phi(k)} \\
P(k) = P(k-1) + R_1 - K(k)\phi^T(k)P(k-1),
\end{cases}
\tag{4.13}
$$

in which $P(k)$ denotes the covariance of the estimation error and R_1 and R_2 represent the covariance matrices of the process and the measurement noise, respectively. More specifically, the diagonal elements of R_1 shall be chosen in accordance with how quickly the corresponding parameter varies w.r.t. time. For instance, if the parameter varies slowly, the related element of R_1 should be small and vice versa. When there is no process noise (i.e., $R_1 = 0, R_2 = 1$), the above Kalman filter parameter identification algorithm is equivalent to the recursive least square method. However, due to its ability to recursively calculate the parameters by combining prior knowledge, predictions from system models and noisy measurements, the Kalman filter-based parameter identification algorithm is utilized in this study [63].

Model Predictive Controller Development

As one of the most popular optimal control approaches, the MPC algorithm has been extensively applied both in simulation and real-world application because of its many advantages [64–67]. It was first developed and applied in the chemistry industry, which is featured by its slow dynamics and thus has enough time to solve the optimization problem. As aforementioned, the smart charger with the slow dynamics of the battery SOC makes it appropriate to apply the MPC algorithm. The objective function in this optimal control problem is presented below:

$$J\left(x_0, u_0\right) = \left(y(N) - y_{ref}\right)^T P\left(y(N) - y_{ref}\right)$$
$$+ \sum_{k=0}^{N-1} \left[\left(y(k) - y_{ref}\right)^T Q\left(y(k) - y_{ref}\right) + \left(\frac{u(k)}{\eta(k)}\right)^T R\left(\frac{u(k)}{\eta(k)}\right)\right] \tag{4.14}$$

$s.t.$

$$y_{\min} \le y(k) \le y_{\max}, \ k = 0, \ldots N - 1$$
$$u_{\min} \le u(k) \le u_{\max}, \ k = 0, \ldots N - 1.$$

On the right side of Equation (4.14), the first term denotes the terminal cost, while the second one means a combination of the process costs and the additional fuel consumed to charge the battery, which ensures the direct charging of the battery by the engine, during the high engine-efficiency region. P, Q, and R are the normalized weighting factors for each term. The original objective function can be reconstructed and represented by a quadratic form only w.r.t. the control input. Because the prediction horizon length is chosen to be N, the trajectory of the predicted states and the outputs will be acquired by the discretized system model:

$$\underbrace{\begin{bmatrix} x(k+1) \\ x(k+2) \\ \vdots \\ x(k+N) \end{bmatrix}}_{\bar{X}} = \underbrace{\begin{bmatrix} A \\ A^2 \\ \vdots \\ A^N \end{bmatrix}}_{S^x} x(k) + \underbrace{\begin{bmatrix} B_u & 0 & \cdots & 0 \\ AB_u & B_u & \cdots & 0 \\ \vdots & \vdots & \ddots & 0 \\ A^{N-1}B_u & A^{N-2}B_u & \cdots & B_u \end{bmatrix}}_{S^u} \underbrace{\begin{bmatrix} u(k) \\ u(k+1) \\ \vdots \\ u(k+N-1) \end{bmatrix}}_{\bar{U}}$$

$$+ \underbrace{\begin{bmatrix} B_v(k) \\ B_v(k) + B_v(k+1) \\ \vdots \\ B_v(k) + B_v(k+1) + \ldots B_v(k+N-1) \end{bmatrix}}_{\bar{V}} \tag{4.15}$$

$$\underbrace{\begin{bmatrix} y(k+1) \\ y(k+2) \\ \vdots \\ y(k+N) \end{bmatrix}}_{\bar{Y}} = \underbrace{\begin{bmatrix} C & 0 & 0 & 0 \\ 0 & C & 0 & 0 \\ 0 & 0 & \ddots & 0 \\ 0 & 0 & 0 & C \end{bmatrix}}_{C^x} \bar{X}.$$

The quadratic form of the objective function is obtained, where the only unknown to be calculated is input as follows:

$$J(x_0, u_0) = \frac{1}{2}\bar{U}^T H \bar{U} + \bar{U}^T g$$

$$H = 2(C^x S^u)^T \bar{Q}(C^x S^u) + (1/\eta)^T \bar{R}(1/\eta), \quad g = 2(C^x S^u)^T \bar{Q}(C^x S^u - \bar{Y}_{ref})$$

s.t.
$$\tag{4.16}$$

$$\bar{U} \geq \max\left(\bar{U}_{\min}(U), \bar{U}_{\min}(\bar{U}), \bar{U}_{\min}(X)\right)$$
$$\bar{U} \leq \min\left(\bar{U}_{\max}(U), \bar{U}_{\max}(\bar{U}), \bar{U}_{\max}(X)\right),$$

where the matrix (H) denotes symmetric and semi-positive or positive definite while g means the gradient vector. $\bar{Q}, \bar{R}$, and $\bar{X}_{ref}$ is reconstructed based on the dimensions of the system, N, Q, R, and X_{ref}. The corresponding constraints should also be updated accordingly.

Figure 4.6 shows a predefined (i.e., normal) drive cycle and a real-world one. Because of the aforementioned issues, the real drive cycle cannot be identical to the predefined one but locates around it. However, the real driving information in the near future is actually not accessible as *a priori*, and the only information available is the predefined or the nominal drive cycle. If the prediction step size (i.e., the distance between two prediction points in time scale) is increased to a large number when designing the MPC, the effects on state-trajectory prediction of both drive cycles are closed to each other. The reason is that both the positive and negative deviations around the nominal one counteract each other's influence on the change of *SOC*. In addition, more accurate future information is utilized in the MPC design such that the obtained solution will be much closer to the global optimal ones but without adding any additional computational efforts. To show the advantages brought upon by the proposed MPC, the prediction step size (i.e., 10 s) is selected. In other words, the future states are predicted every 10 s by Equation (4.17), and then used in the updated output trajectory given in Equation (4.15) to form the quadratic problem (QP):

$$
\underbrace{\begin{bmatrix} y(k+10) \\ y(k+20) \\ \vdots \\ y(k+N*10) \end{bmatrix}}_{\bar{Y}} = \underbrace{\begin{bmatrix} C & 0 & 0 & 0 \\ 0 & C & 0 & 0 \\ 0 & 0 & \ddots & 0 \\ 0 & 0 & 0 & C \end{bmatrix}}_{C^x} \begin{bmatrix} x(k+10) \\ x(k+20) \\ \vdots \\ x(k+N*10) \end{bmatrix}.
\tag{4.17}
$$

Nevertheless, once the optimal solutions are found, the first element will still be sent to and control the system for the next step, and then the algorithm utilizes the fresh data to repeat.

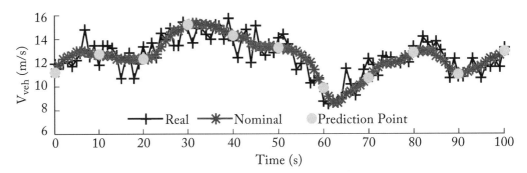

Figure 4.6: Drive cycles and prediction points of the proposed MPC.

Service Cycle

The service cycle or duty cycle can be identified by the method mentioned in previous sections. However, for the sake of simplicity and to show the PMSs design process as well as the preliminary results, this section creates a service cycle based on the ambient conditions. As the accessory that consumes the main part of the auxiliary power, A/C-R unit operating power is related to the ambient and operating conditions. Figure 4.7 indicates the ambient air temperature and the thermal load exerted onto the cargo from 10:00 am to 12:30 pm on a summer morning. Simply, the thermal load is considered to be linearly related to the deviation between the outside and inside temperature of the vehicle.

Some additional thermal load exists during the loading/unloading periods when the door is open. This additional load is roughly assumed to be 0.1 kW, as shown in Figure 4.8, where

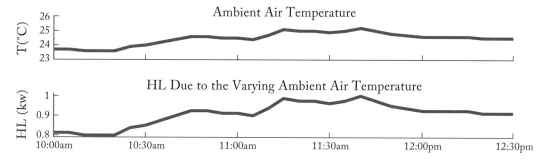

Figure 4.7: The ambient temperature and corresponding HL.

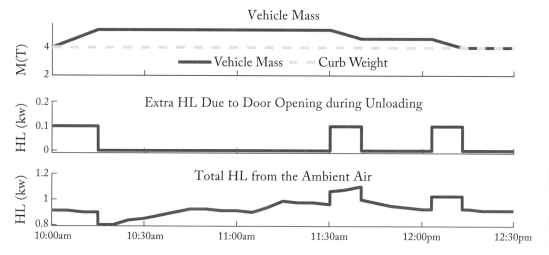

Figure 4.8: Vehicle mass, extra HL, and total HL.

the total ambient thermal load is calculated by the method used in [68]. The cooling capacity generated by the A/C-R device is only used to offset the total thermal load to keep the required temperature. The ratio of the power consumed by the A/C-R unit to the cooling capacity generated in this study is set to 1.

Drive Cycle

The nominal or the predefined driving cycle is generated in accordance with the segment provided in the first column of Table 4.3.

Two types of disturbances are applied on the nominal drive cycle to simulate the real-world. In the first scenario, about 15% white noise is given and the locations of two segments

Table 4.3: Drive cycles

Nominal		Scenario 1	Scenario 2	
Segment	Duration(s)		Segment	
Loading	900		Unloading	
FTP75	1874		FTP75	
HWFET	765		HWFET	
FTP75	1874		UDDS	
	Nominal + Disturbance		Switch the orders of two segments + disturbance	
Unloading	600		Unloading	
UDDS	1370		FTP75	
Unloading	600		Unloading	
UDDS	1017		UDDS	

are switched to form the drive cycle in scenario 2. The drive cycles of 2.5 h are visualized in Figure 4.9, where the middle subfigure indicates the amplified velocity around 10:30 am.

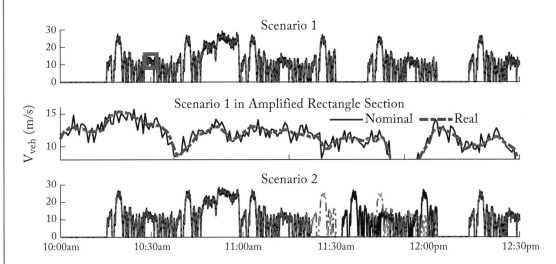

Figure 4.9: Drive cycles under two scenarios.

Drive Cycle and Service Cycle Analysis

In this study, the studied vehicle is the GMC SAVANA 2500. Take the UDDS drive cycle for instance, according to vehicle longitudinal dynamics mentioned in Chapter 2. The vehicle power is calculated and shown in Figure 4.10, where only the power for driving the vehicle and its average value is presented.

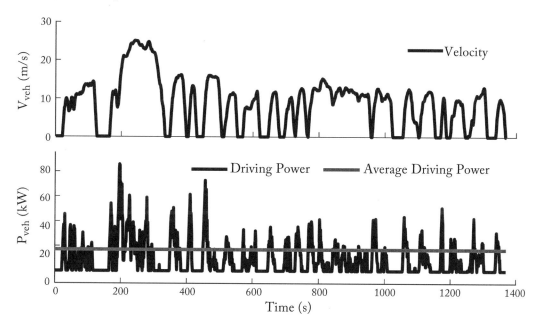

Figure 4.10: Drive cycles of UDDS and vehicle power.

Figure 4.8 shows that the maximum auxiliary power is 1.1 kW. Compared to the driving power of the vehicle, the auxiliary power is relatively small. Thus, this kind of vehicle refers to the light service vehicle, and the average auxiliary and driving power are provided in Table 4.4.

Table 4.4: Driving power and auxiliary power

Drive Cycle	UDDS	The Proposed Drive Cycle
Average driving (auxiliary) power (kW)	17.15 (0.9)	21.53 (0.9)

MPCs Configuration and Results Analysis

Once the drive and service cycle are available, such information is fed to the MPC to find the optimal solution. In this section, MPC[1] denotes the prescient MPC, which has access to the real drive cycle beforehand and can be used as a benchmark to evaluate other methods, whereas

MPC^2 only has access to the nominal drive cycle. The parameters of the controllers are indicated in Table 4.5.

Table 4.5: MPC parameters

Parameters	T_s (sec)	N	Q	P	R	Y_{ref}
Value	10/1	10	$\begin{bmatrix} 0.5 & 0 \\ 0 & 1 \end{bmatrix}$	$10Q$	1	$\begin{bmatrix} 0.9 \\ 0 \end{bmatrix}$

To properly use the battery, the *SOC* should not be depleted and reach its bounds to ensure high overall efficiency [69–71]. Since the battery powers the accessory device to eliminate engine idling, the battery *SOC* should be kept at a higher level before vehicle stops. Therefore, 0.9 is selected as the *SOC* reference and a small weight 0.5 is selected to avoid it deviating from its reference too far. The prediction horizontal length N must be chosen properly to balance the optimality and time consumed. A relatively large terminal weight is adopted to improve the stability of the controller. An open-source solver, capable of solving the QP in milliseconds is used to solve the QP [72].

The responses of the inputs, *SOC*, and vehicle mass estimation during the whole process are shown in Figures 4.11, 4.12, 4.13, 4.14, 4.15, and 4.16. Both scenarios conclude that MPC^1 used less energy to charge the battery leading to less fuel consumption. Although MPC^2 utilizes the nominal one to forecast the future state response of the system, it can still perform similarly as MPC^1. As expected, the MPC with a large step size outperforms the other because such MPC can predict over a longer horizon, offset the influence of the disturbances and react much earlier. The Kalman filter can predict the mass exactly as long as the vehicle does not stop.

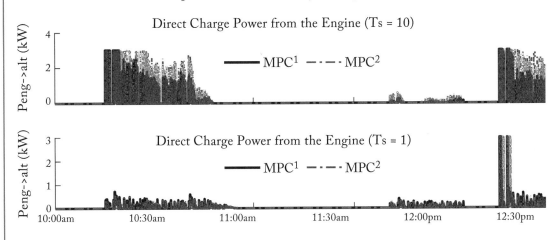

Figure 4.11: System inputs under scenario 1.

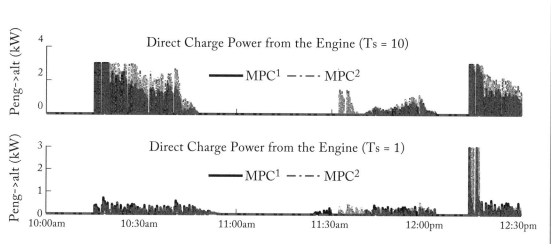

Figure 4.12: System inputs under scenario 2.

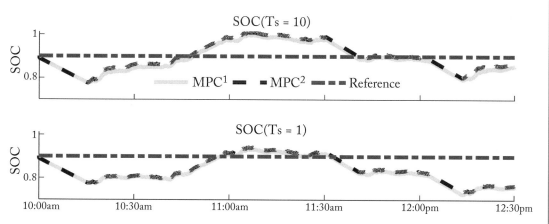

Figure 4.13: *SOC* responses under scenario 1.

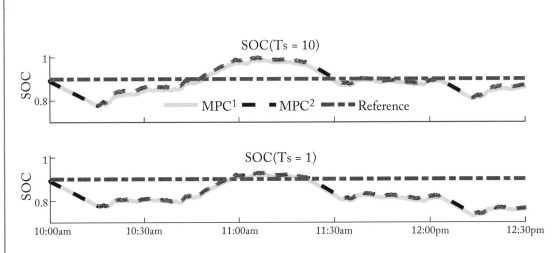

Figure 4.14: *SOC* responses under scenario 2.

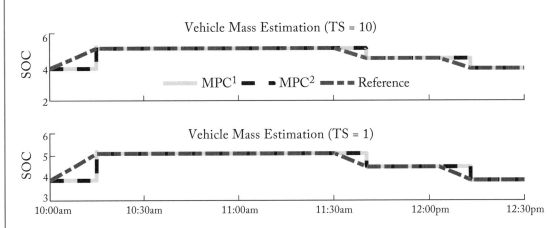

Figure 4.15: Vehicle mass estimation under scenario 1.

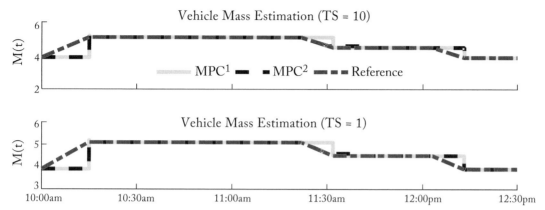

Figure 4.16: Vehicle mass estimation under scenario 2.

Table 4.6: Result comparisons for both scenarios

Systems		Scenario 1					Scenario 2			
		Terminal SoC	FC¹ (L)	Save¹ (%)	FC² (L)	Save² (%)	Final SOC	FC¹ (L)	Save¹ (%)	FC² (L)	Save² (%)
Without RAPS		N/A	19.7	Basis	1.56	Basis	N/A	19.7	Basis	1.56	Basis
With RAPS	MPC¹ (T_s=10)	0.8649	18.273	7.24	0.133	91.46	0.8722	18.279	7.21	0.139	91.05
	MPC² (T_s=10)	0.8782	18.277	7.22	0.137	91.19	0.8817	18.309	7.06	0.169	90.14
	MPC¹ (T_s=1)	0.7689	18.302	7.09	0.162	89.58	0.7708	18.308	7.07	0.168	89.26
	MPC² (T_s=10)	0.7756	18.31	7.06	0.17	89.13	0.7756	18.412	6.54	0.272	82.53
Vehicle		N/A	18.14	N/A	N/A	N/A	N/A	18.14	N/A	N/A	N/A

FC¹: total FC; FC²: FC of A/C-R system; Save¹: fuel saving percentage due to adding the smart charger; Save²: fuel saving percentage of A/C-R unit

Table 4.6 presents the results of the terminal *SOC* and the consumed fuel. In order to eliminate the influence of the deviation of the terminal *SOCs* on the fuel consumption, a *SOC*-correction method [73] is employed. The totally consumed fuel (19.7 L) for the vehicle without the smart charger is provided as a benchmark. Meanwhile, the fuel (18.14 L) only used to drive vehicle, therefore the difference is the fuel for powering the A/C-R unit. In scenario 1, about 7% savings are witnessed in "Save¹" column once the smart charger is added to the conventional powertrain. Furthermore, the MPCs with a larger step size save more fuel. As for FC^2 and $Save^2$, the fuel consumed to power the auxiliary power system in the conventional vehicle is 1.56 L, whereas the fuel used in the vehicles with the smart charger is nine times less since the recovered energy is also used to power the auxiliary system. Scenario 2 can draw a similar conclusion as well. However, in this case, the drive cycle adopted for prediction seems totally different from the real one, therefore, the savings are less than that of the scenario 1. It indicates the proposed MPC is robust to such large disturbances.

Remark: In the current study, the average auxiliary power is about 1 kW, which is relatively smaller than the power to drive the vehicle. Therefore, such vehicles are referred to as light service vehicles. The above results are obtained based on a light service vehicle with only 10% braking power captured via a serpentine belt. For the heavy service vehicle with a PTO configuration, more fuel savings will be witnessed by recovering more braking power.

4.2.2   AVERAGE-CONCEPT BASED MPC

This section presents a model predictive PMS for the smart charger used in service vehicles, which do not know the future drive information *a priori*. As one of the most popular optimal control methods, the MPC requires the drivers' command or the drive cycle at least in the near future to be available beforehand. However, in this research, an average-concept-based MPC is designed to roughly identify such information, as discussed in Section 3.2. The analysis shows that the smart charger with the proposed MPC saves fuel. At the same time, this proposed MPC performs similarly with the prescient MPC. Moreover, its robustness is also verified under other scenarios. The presented MPC does not rely on the powertrain topology so it can be easily extended to other kinds of powertrains. More importantly, it shows a statistic option to use MPC when there is no direct access to future driving formation [73].

Model Predictive PMS Development
The same procedure is used as discussed in the previous section. Based on that, several MPCs are developed with different acquisition methods of future information. In the following table, three MPCs are shown and compared afterward. Since the estimated future information is a direct reflection of the historical data, it cannot be used exactly but in an average sense. Therefore, it is not correct to identify when the engine runs in the high efficiency region in the prediction horizon based on the estimated information. That is why an additional rule is considered to check if the result by MPC^2 is good enough to be used. In other words, MPC^2 won't send the solution directly to the plant once obtaining it. Instead, it calculates the current engine efficiency meanwhile. The calculated optimal control input won't be used unless the efficiency is higher than a lower limit; otherwise, a zero control action is sent out. In this way, the present one ensures to directly charge the battery in a high engine-efficiency period.

Table 4.7: Types of the model predictive controllers

MPC Types	Description
MPC^1	Prescient MPC
MPC^2	Utilizes the latest 100 s driving info as the future driving info
MPC^3	MPC^2 +a rule

Case Study

In the present section, all the aforementioned control strategies are studied in three cases. First, the three MPCs are evaluated under the standard urban driving scenario UDDS, where the controller parameters are tuned. Using these properly tuned parameters, all the MPCs are then verified under a completely different highway driving scenario. Last, all the tuned controllers are evaluated over a more realistic driving scenario for a delivery truck. During the whole process, the service load is kept the same as mentioned in the previous sections.

Urban driving scenario   The controller parameters should be determined or tuned before conducting the simulation experiments. The tuned parameters of the MPCs are provided in Table 4.8 by conducting many simulations. To protect and properly use the battery, its SOC must be kept in a certain range. Furthermore, the battery will be used to power auxiliary devices to reduce idling once the vehicle stops such that its SOC must stay at a high level before the vehicle reaching its stops. Moreover, the SOC shall not touch its bounds to pursue the high efficiency. As a result, the same as before, 0.9 is selected as a reference and a weight (0.5) is chosen to prevent the SOC going far from its set point. According to the rule to choose the weights in the previous section, the control parameters are presented in Table 4.8.

Table 4.8: MPC parameters

Parameters	T_s (sec)	N	Q	P	R	Y_{ref}
Value	10	10	0.5	10Q	1	0.9

The SOC response of each controller is shown in Figure 4.17, where SOC starts to change from the same initial value of 0.9.

The prescient MPC (i.e., MPC^1) charges the battery not as much as others since the SOC is still far from its bound 0.6 and it only charges the battery if needed in the periods when the engine is running in its high-efficiency region, as indicated in Figure 4.18.

The MPC^2 uses the latest historical data as the near future info and recharges the battery more in both high and low engine-efficiency periods. If a lower efficiency threshold is intro-

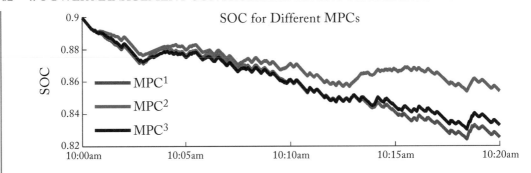

Figure 4.17: *SOC* performance of different MPC.

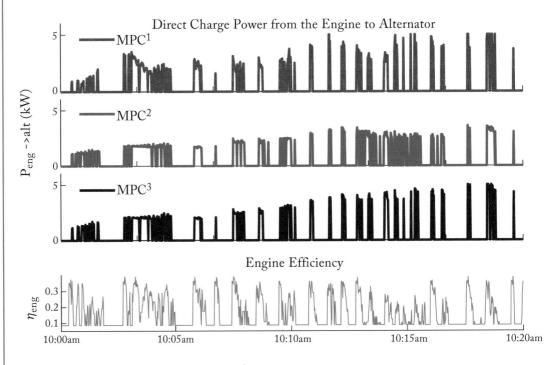

Figure 4.18: System input and engine efficiency.

duced, the MPC[3] only charges the battery if the current engine efficiency is higher than this threshold.

In order to show the benefits of the designed smart charger as well as the controller in terms of the fuel efficiency, the consumed fuel in several scenarios is presented in Table 4.9. The fuel used by the conventional powertrain is 2.083 L and the fuel for powering the vehicle is 1.755 L. Because the smart charger is adopted for electrifying the accessory device (e.g., A/C-R unit), it does not affect the amount of fuel for powering the vehicle. Yet, it is able to minimize the consumed fuel by the A/C-R device in an optimal manner. In addition, the same correction method is used to compensate the difference between the initial and terminal SOC such that the comparison is fair. After adding the smart charger into the conventional powertrain, it sees a significant saving under the studied scenarios. However, the saving percentages depend on which MPC is used. Even though MPC^2 saves less than MPC^1 after the efficiency threshold is added, MPC^3 performs as well as MPC^1 does. The engine efficiency threshold (15%) in MPC^3 is tuned under this case and will be adopted in other scenarios.

Table 4.9: Result comparisons for each controller

		Terminal SoC	FC^1 (L)	Save^1 (%)	FC^2 (L)	Save^2 (%)
Without the smart charger		N/A	2.083	Basis	0.328	Basis
With the smart charger	MPC^1	0.8251	1.841	9.22	0.136	58.55
	MPC^2	0.8536	1.912	6.72	0.188	42.70
	MPC^3	0.8326	1.849	9.08	0.139	57.63
Vehicle		N/A	1.755	N/A	N/A	N/A

FC^1: total fuel consumed; FC^2: fuel consumed by A/C-R system; Save^1: fuel saving because of adding the smart charger; Save^2: fuel saving by A/C-R system thanks to adding the smart charger

Highway scenario (HWFET+US06)   To evaluate the reliability of the tuned parameters in the previous section as well as the robustness of the designed MPC, a highway driving scenario is created, including two classical highway cycles. According to Figures 4.19 and 4.20 and Table 4.10, the highway study can draw similar conclusions to that of the previous city drive cycle UDDS, but with less fuel savings.

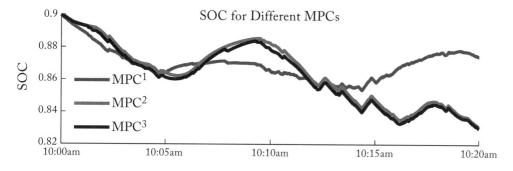

Figure 4.19: SOC performance of different MPC.

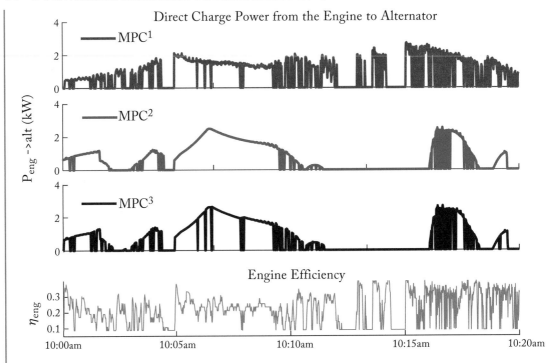

Figure 4.20: The results of the system input and engine efficiency.

Table 4.10: Result comparisons for each controller

		Terminal SoC	FC[1] (L)	Save[1] (%)	FC[2] (L)	Save[2] (%)
Without the smart charger		N/A	4.240	Basis	0.278	Basis
With the smart charger	MPC[1]	0.8744	4.158	1.93	0.153	34.86
	MPC[2]	0.8308	4.178	1.46	0.173	26.31
	MPC[3]	0.8298	4.165	1.77	0.160	31.99
Vehicle		N/A	4.005	N/A	N/A	N/A

The reason is because unlike the urban driving, highway driving cannot recollect braking power as frequently and as much. Still, it can be concluded that even though the future driving information employed for predictions comes from the historical data, MPC[3] can perform similarly with MPC[1] in a very different scenario. Therefore, the proposed MPC is also robust to the driving conditions.

Combined driving scenario This driving scenario is generated to study the performance of the proposed controller in a more realistic driving schedule of delivery or long-haul trucks. The combined driving scenario consists of different conditions, such as highway driving, urban driving, and loading/unloading stops. The specific feature of the combined drive scenario for 2.5 h duration is provided in Table 4.11 and Figure 4.21.

Table 4.11: The combined driving cycle

Segment	FTP75	HWFET	Resting	UDDS	FTP75	US06	Resting	UDDS
Duration(s)	1874	765	600	1370	1874	600	600	1017

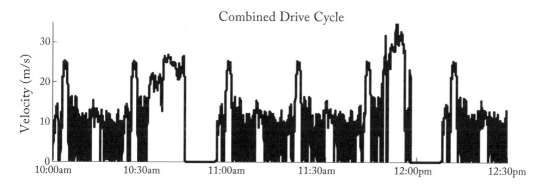

Figure 4.21: Combined drive cycle.

All the developed MPCs are studied over the combined driving scenario with the exact parameters as the previous two sections did. As the only system state, *SOC*'s performance and the input response are presented in Figures 4.22 and 4.23. It can be witnessed that even under a much different and complex driving cycle and without any available future driving information, MPC³ still presents a similar performance as MPC¹. As a result, it can be concluded that the developed, statistic-based MPC³ can be used in the proposed smart charger and other hybrid powertrains as well, with a sub-optimal performance.

Remarks: On one hand, as stated above, the auxiliary power consumed by the A/C-R unit in this studied delivery truck is about 1 kW, making it a light duty truck. Nevertheless, the smart charge system can still reduce about 10% in fuel consumption. Regarding the heavy-duty vehicles, more benefits should be witnessed.

On the other hand, the simulation studies above are conducted in the charge and sustaining mode [74]. In fact, if the charging station and destination knowledge are accessible from vehicle to infrastructure (V2I) information, the controller for the charge and depleting mode [75] can also be applied by easily extending the algorithm. The idea behind this is that according to the information—such as the available regenerative braking, the auxiliary demand power, the *SOC* status, and the time or the distance left to the charging stations—the developed

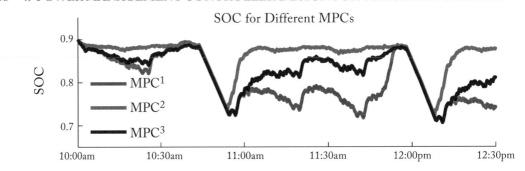

Figure 4.22: *SOC* performance for each controller.

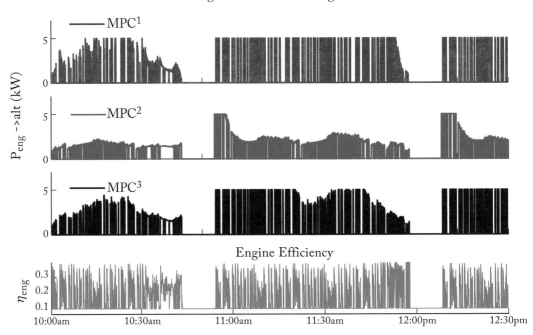

Figure 4.23: The responses of the system input and engine efficiency.

controller will be able to determine if charging the battery is beneficial and ensure that the battery completely depletes when reaching the stations or destination, where the plug-in energy is available.

Table 4.12: Result comparisons for each controller

		Terminal SoC	FC^1 (L)	Save^1 (%)	FC^2 (L)	Save^2 (%)
Without the smart charger		N/A	16.3941	Basis	2.563	Basis
With the smart charger	MPC^1	0.7414	14.955	8.75	1.126	56.02
	MPC^2	0.8779	15.934	2.78	2.105	17.78
	MPC^3	0.8106	14.990	8.54	1.160	54.70
Vehicle		N/A	13.831	N/A	N/A	N/A

4.3   SUMMARY

The aim of this section was to design a robust predictive PMS for the developed smart charger when the future driving information is inaccessible. By adding the proposed smart charger into the conventional powertrain of a delivery truck or other service vehicles, the accessory devices are electrified and the related idling is eliminated. The smart charge system together with different MPCs was studied by simulations under several scenarios. The first scenario was studied to determine and tune the controller parameters whereas the others were employed to evaluate the effectiveness and robustness of each designed controller. The simulation results showed around 10% improvement in fuel economy could be experienced by adding the proposed smart charger and the corresponding MPCs in the light-duty service vehicles under all cases. Then, the designed MPC^3 is able to figure out the sub-optimal solution much closer to the optimal one even when the future driving information is inaccessible. Above all, the results indicated that the proposed controller is effective and robust enough to apply to any drive and load conditions.

CHAPTER 5

Conclusions

The ultimate goal of this book was to design and realize a robust controller for the smart charger to reduce engine idling for service vehicles (e.g., a delivery truck studied in this research) with A/C-R systems as the major auxiliary devices.

Two potential configurations for the smart charger were proposed for heavy- and light-duty service vehicles by connecting the generator to the serpentine belt or a PTO to recover the braking energy to charge the battery. To maximize the benefits of the smart charger system, it should be optimized at a system level. Since there are different parts with different disciplines in the smart charger, a Multi-disciplinary Design Optimization (MDO) technique should be utilized. The first step was to develop a scalable and composable system model that could be used by the optimizer algorithm. The MDO optimizer simultaneously optimizes the component sizing and power management logic in the overall system.

Information, such as the auxiliary demand power and the future driving cycle, is necessary when developing advanced PMSs (e.g., MPC). Therefore, Chapter 3 introduced two prediction methods for the auxiliary load and the driving cycle. Both methods work by using the historical data collected by the onboard sensors. All the predicted results were used for designing PMSs. Two types of real-time PMSs (i.e., AECMS and MPC) were developed and several case studies were conducted to demonstrate their effectiveness, robustness, and benefits in terms of the fuel saving. From the results, it can be seen the smart charger can save about 10% fuel if it is added to the conventional powertrain.

References

[1] M. Morshed, Unnecessary idling of vehicles: An analysis of the current situation and what can be done to reduce it, M.S. thesis, McMaster University, Hamilton, ON, Canada, 2010. 1

[2] M. Pournazeri, A. Khajepour, and Y. Huang, Development of a new fully flexible hydraulic variable valve actuation system for engines using rotary spool valves, *Mechatronics*, vol. 46, pp. 1–20, 2017. DOI: 10.1016/j.mechatronics.2017.06.010. 1

[3] Y. Huang, *Anti-idling systems for service vehicles with A/C-R units: Modeling, holistic control, and experiments*, Ph.D. dissertation, University of Waterloo, Waterloo, ON, Canada, 2016. 1, 2, 4

[4] Z. Han, N. Xu, H. Chen, et al., Energy-efficient control of electric vehicles based on linear quadratic regulator and phase plane analysis, *Applied Energy*, vol. 213, pp. 639–657, 2018. DOI: 10.1016/j.apenergy.2017.09.006. 2

[5] M. Hua, G. Chen, B. Zhang, et al., A hierarchical energy efficiency optimization control strategy for distributed drive electric vehicles, *Proc. of the Institution of Mechanical Engineers, Part D: Journal of Automobile Engineering*, p. 095440701775178, 2018. DOI: 10.1177/0954407017751788.

[6] C. Hu, R. Wang, F. Yan, et al., Differential steering based yaw stabilization using ISMC for independently actuated electric vehicles, *IEEE Transactions on Intelligent Transportation Systems*, vol. 19, no. 2, pp. 627–638, 2018. DOI: 10.1109/tits.2017.2750063. 2

[7] V. Madanipour, M. Montazeri-Gh, and K. M. Mahmoodi, Multi-objective component sizing of plug-in hybrid electric vehicle for optimal energy management, *Clean Technologies and Environmental Policy*, vol. 18, no. 4, pp. 1–14, 2016. DOI: 10.1007/s10098-016-1115-1. 2

[8] X. Tang, X. Hu, W. Yang, and H. Yu, Novel torsional vibration modeling and assessment of a power-split hybrid electric vehicle equipped with a dual-mass flywheel, *IEEE Transactions on Vehicular Technology*, vol. 67, no. 3, pp. 1990–2000, 2018. DOI: 10.1109/tvt.2017.2769084. 2

[9] M. Khazraee, Y. Huang, and A. Khajepour, Anti-idling systems for service vehicles: Modeling and experiments, *Proc. of the Institution of Mechanical Engineers,*

Part K: Journal of Multi-body Dynamics, vol. 232, no. 1, pp. 49–68, 2017. DOI: 10.1177/1464419317709397. 2, 7, 11

[10] Y. Huang, A. Khajepour, T. Zhu, and H. Ding, A supervisory energy-saving controller for a novel anti-idling system of service vehicles, *IEEE/ASME Transactions on Mechatronics*, vol. 22, no. 2, pp. 1037–1046, 2017. DOI: 10.1109/tmech.2016.2631897. 2

[11] Y. Wang, et al., Investigating the cost, liability and reliability of anti-idling equipment for trucks, *Delaware Department of Transportation*, Report, 2007. 2, 5

[12] P. Kuo, Evaluation of freight truck anti-idling strategies for reduction of greenhouse gas emissions, Ph.D. dissertation, North Carolina State University, Raleigh, NC, 2008.

[13] H. Christopher and P. Kuo, Best practices guidebook for greenhouse gas reductions in freight transportation, *U.S. Department of Transportation*, Report, 2007. 2

[14] Y. Huang, M. Khazeraee, H. Wang, S. Fard, T. Zhu, and A. Khajepour, Design of a regenerative auxiliary power system for service vehicles, *Automotive Innovation*, vol. 1, no. 1, pp. 62–69, 2018. DOI: 10.1007/s42154-018-0008-x. 4

[15] http://www.mass.gov/eea/docs/doer/clean-cities/air-dock.pdf 4, 5

[16] G. Gereffi and K. Dubay, Auxiliary power units: Reducing carbon emissions by eliminating idling in heavy-duty trucks, *Duke CGGC*, Report, 2008. 4

[17] http://idlefreesystems.com/no-idle-elimination-systems-battery.html 4

[18] http://www.ecamion.com/portfolio/eapu/,Accessed:2018--04-11 4

[19] E. Lust, System-level analysis and comparison of long-haul truck idle-reduction technologies, M.S. thesis, University of Maryland, College Park, MD, 2008. 4

[20] http://www.eaton.com/Eaton/ProductsServices/HybridPower/index.htm 4

[21] http://www.freightlinertrucks.com/Trucks/Alternative-Power-Trucks/Hybrid-Electric 4

[22] http://www.mitsubishifuso.com/en/products/truck/canter_hev/12/concepts/environment/hybrid/index.html 4

[23] Michael A. Tunnell, et al., Idle reduction technology: Fleet preferences survey, *New York State Energy Research and Development*, Report, 2006. 5

[24] Government of Alberta, Draft quantification protocol for reduced vehicle idling through the use of direct fired heaters, *Climate Change Secretariat*, Edmonton, AB, Canada, 2011. 5

[25] Y. Huang, A. Khajepour, F. Bagheri, and M. Bahrami, Modelling and optimal energy-saving control of automotive air-conditioning and refrigeration systems, *Proc. of the Institution of Mechanical Engineers, Part D: Journal of Automobile Engineering*, vol. 231, no. 3, pp. 291–309, 2016. DOI: 10.1177/0954407016636978. 5

[26] Y. Huang, A. Khajepour, H. Ding, F. Bagheri, and M. Bahrami, An energy-saving set-point optimizer with a sliding mode controller for automotive air-conditioning/refrigeration systems, *Applied Energy*, vol. 188, pp. 576–585, 2017. DOI: 10.1016/j.apenergy.2016.12.033. 5

[27] B. S. Fan, *Multidisciplinary Optimization of Hybrid Electric Vehicles: Component Sizing and Power Management Logic*, University of Waterloo, 2011. 9

[28] L. Guzzella and C. H. Onder, *Introduction to Modeling and Control of Internal Combustion Engine Systems*, Springer, 2009. DOI: 10.1007/978-3-642-10775-7. 10

[29] Y. Qin, C. He, X. Shao, H. Du, C. Xiang, and M. Dong, Vibration mitigation for in-wheel switched reluctance motor driven electric vehicle with dynamic vibration absorbing structures, *Journal of Sound and Vibration*, vol. 419, pp. 249–267, 2018. DOI: 10.1016/j.jsv.2018.01.010. 11

[30] X. Tang, W. Yang, D. Zhang, and H. Yu, A novel two degree of freedom model of a full hybrid electric vehicle, *International Journal of Electric and Hybrid Vehicles*, vol. 9, no. 1, pp. 67–77, 2017. DOI: 10.1504/IJEHV.2017.082817. 11

[31] H. Wang, Y. Huang, A. Khajepour, and Q. Song, Model predictive control-based energy management strategy for a series hybrid electric tracked vehicle, *Applied Energy*, vol. 182, pp. 105–114, 2016. DOI: 10.1016/j.apenergy.2016.08.085. 11

[32] H. He, R. Xiong, H. Guo, and S. Li, Comparison study on the battery models used for the energy management of batteries in electric vehicles, *Energy Conversion and Management*, vol. 64, no. 0, pp. 113–121, 2012. DOI: 10.1016/j.enconman.2012.04.014. 13

[33] M. Pournazeri, A. Khajepour, and Y. Huang, Improving energy efficiency and robustness of a novel variable valve actuation system for engines, *Mechatronics*, vol. 50, pp. 121–133, 2018. DOI: 10.1016/j.mechatronics.2018.02.002. 13

[34] K. Liu, Y. Li, J. Yang, et al., Comprehensive study of key operating parameters on combustion characteristics of butanol-gasoline blends in a high speed SI engine, *Applied Energy*, vol. 212, pp. 13–32, 2018. DOI: 10.1016/j.apenergy.2017.12.011. 13

[35] G. Rizzoni, L. Guzzella, and B. M. Baumann, Unified modeling of hybrid electric vehicle drivetrains, *IEEE/ASME Transactions on Mechatronics*, vol. 4, no. 3, pp. 246–257, 1999. DOI: 10.1109/3516.789683. 13

[36] Y. Qin, R. Langari, Z. Wang, C. Xiang, and M. Dong, Road excitation classification for semi-active suspension system with deep neural networks, *Journal of Intelligent and Fuzzy Systems*, vol. 33, no. 3, pp. 1907–1918, 2017. DOI: 10.3233/jifs-161860. 14

[37] Y. Qin, M. Dong, R. Langari, L. Gu, and J. Guan, Adaptive hybrid control of vehicle semiactive suspension based on road profile estimation, *Shock and Vibration*, 14 Article ID: 636739, 2015. DOI: 10.1155/2015/636739.

[38] N. Shidore, J. Kwon, and A. Vyas, Trade-off between PHEV fuel efficiency and estimated battery cycle life with cost analysis, in *IEEE Vehicle Power and Propulsion Conference*, no. 978, pp. 669–677, 2009. DOI: 10.1109/vppc.2009.5289784. 16

[39] S. Fard, Y. Huang, M. Khazraee, and A. Khajepour, A novel anti-idling system for service vehicles, *Energy*, vol. 127, pp. 650–659, 2017. DOI: 10.1016/j.energy.2017.04.018. 29

[40] J. MacQueen, Some methods for classification and analysis of multivariate observations, *Proc. of the 5th Berkeley Symposium on Mathematical Statistics and Probability*, vol. 1, p. 281–97, Oakland, CA, 1967. 30

[41] Y. Qin, R. Langari, and L. Gu, A new modeling algorithm based on ANFIS and GMDH, *Journal of Intelligent and Fuzzy Systems*, vol. 29, no. 4, pp. 1321–1329, 2015. DOI: 10.3233/ifs-141443. 30

[42] M. Ester, H.-P. Kriegel, J. Sander, and X. Xu, A density-based algorithm for discovering clusters in large spatial databases with noise, *KDD*, vol. 96, pp. 226–231, 1996. 30, 32

[43] Y. Huang, A. Khajepour, and H. Wang, A predictive power management controller for service vehicle anti-idling systems without a priori information, *Applied Energy*, vol. 182, pp. 548–557, 2016. DOI: 10.1016/j.apenergy.2016.08.143. 34

[44] Y. Huang, H. Wang, A. Khajepour, H. He, and J. Ji, Model predictive control power management strategies for HEVs: A review, *Journal of Power Sources*, vol. 341, pp. 91–106, 2017. DOI: 10.1016/j.jpowsour.2016.11.106. 34, 48

[45] H. Banvait, J. Hu, and Y. Chen, Energy management control of plug-in hybrid electric vehicle using hybrid dynamical systems, in *IEEE Transactions on Intelligent Transportation Systems*, 2013. 36

[46] S. Di Cairano, D. Bernardini, A. Bemporad, and I. V. Kolmanovsky, Stochastic MPC with learning for driver-predictive vehicle control and its application to HEV energy management, *IEEE Transactions on Control Systems and Technology*, vol. 22, no. 3, pp. 1018–1031, May 2014. DOI: 10.1109/tcst.2013.2272179.

[47] H. Borhan, A. Vahidi, A. M. Phillips, M. L. Kuang, I. V. Kolmanovsky, and S. Di Cairano, MPC-based energy management of a power-split hybrid electric vehicle, *IEEE Transactions on Control Systems Technology*, vol. 20, no. 3, pp. 593–603, May 2012. DOI: 10.1109/tcst.2011.2134852.

[48] P. Zhang, F. Yan, and C. Du, A comprehensive analysis of energy management strategies for hybrid electric vehicles based on bibliometrics, *Renewable Sustainable Energy Review*, 48, pp. 88–104, August 2015. DOI: 10.1016/j.rser.2015.03.093.

[49] Q. Gong, Y. Li, and Z.-R. Peng, Trip-based optimal power management of plug-in hybrid electric vehicles, *IEEE Transactions on Vehicular Technology*, vol. 57, no. 6, pp. 3393–3401, November 2008. DOI: 10.1109/tvt.2008.921622.

[50] Y. L. Murphey, J. Park, Z. Chen, M. L. Kuang, M. A. Masrur, and A. M. Phillips, Intelligent hybrid vehicle power control Part I: Machine learning of optimal vehicle power, *IEEE Transactions on Vehicular Technology*, vol. 61, no. 8, pp. 3519–3530, October 2012. DOI: 10.1109/tvt.2012.2206064.

[51] C. Sun, X. Hu, S. J. Moura, and F. Sun, Velocity predictors for predictive energy management in hybrid electric vehicles, *IEEE Transactions on Control Systems Technology*, vol. 23, no. 3, pp. 1197–1204, May 2015. DOI: 10.1109/tcst.2014.2359176.

[52] S. Di Cairano, D. Bernardini, A. Bemporad, and I. Kolmanovsky, Stochastic MPC with learning for driver-predictive vehicle control and its application to HEV energy management, *IEEE Transactions on Control Systems Technology*, vol. 22, no. 3, pp. 1018–1031, 2014. DOI: 10.1109/tcst.2013.2272179.

[53] G. Paganelli, S. Delprat, T.-M. Guerra, J. Rimaux, and Santin J.-J., Equivalent consumption minimization strategy for parallel hybrid powertrains, *Vehicle Technology Conference, (VTC), IEEE 55th*, vol. 4, pp. 2076–81, 2002. DOI: 10.1109/vtc.2002.1002989. 40

[54] A. Sciarretta, M. Back, and L. Guzzella, Optimal control of parallel hybrid electric vehicles, *Control System Technology IEEE Transactions*, 12, pp. 352–63, 2004. DOI: 10.1109/tcst.2004.824312. 40

[55] C. Musardo, G. Rizzoni, Y. Guezennec, and B. Staccia, A-ECMS: An adaptive algorithm for hybrid electric vehicle energy management, *European Journal of Control*, 11, pp. 509–524, 2005. DOI: 10.1109/cdc.2005.1582424. 40

[56] S. F. Mohagheghi, and A. Khajepour, An optimal power management system for a regenerative auxiliary power system for delivery refrigerator trucks, *Applied Energy*, 169, pp. 748–756, 2016. DOI: 10.1016/j.apenergy.2016.02.078. 40

[57] D. S. Naidu, *Optimal Control Systems*, CRC Press, 2002. DOI: 10.1115/1.1641776. 43

[58] C. Zhang and A. Vahidi, Route preview in energy management of plug-in hybrid vehicles, *Control System Technology IEEE Transactions on*, 20, pp. 546–53, 2012. DOI: 10.1109/tcst.2011.2115242. 44

[59] A. Vahidi, Route preview in energy management of plug-in hybrid vehicles, *IEEE Transactions Control System Technology*, 20, pp. 546–553, 2012. DOI: 10.1109/tcst.2011.2115242. 44

[60] J. T. B. A. Kessels, M. W. T. Koot, P.P.J. Van den Bosch, and D.B. Kok, Online energy management for hybrid electric vehicles, *Vehicular Technology IEEE Transactions on*, 57, pp. 3428–3440, 2008. DOI: 10.1109/tvt.2008.919988.

[61] S. Fard and A. Khajepour, Concurrent estimation of a vehicle's mass and auxiliary power, *ASME International Mechanical Engineering Congress and Exposition*, vol. 4B, Dynamics, Vibration, and Control, Montreal, Quebec, Canada, 2014. DOI: 10.1115/imece2014-38156. 49

[62] Y. Huang, S. Fard, M. Khazraee, H. Wang, and A. Khajepour, An adaptive model predictive controller for a novel battery-powered anti-idling system of service vehicles, *Energy*, vol. 127, pp. 318–327, 2017. DOI: 10.1016/j.energy.2017.03.119. 49

[63] Y. Qin, R. Langari, Z. Wang, C. Xiang, and M. Dong, Road profile estimation for semi-active suspension using an adaptive Kalman filter and adaptive super-twisting observer, *Proc. of the American Control Conference (ACC)*, Seattle, WA, May 24–26, 2017. DOI: 10.23919/acc.2017.7963079. 50

[64] Y. Huang, A. Khajepour, F. Bagheri, and M. Bahrami, Optimal energy-efficient predictive controllers in automotive air-conditioning/refrigeration systems, *Applied Energy*, vol. 184, pp. 605–618, 2016. DOI: 10.1016/j.apenergy.2016.09.086. 50

[65] J. Ji, A. Khajepour, W. Melek, et al., Path planning and tracking for vehicle collision avoidance based on model predictive control with multiconstraints, *IEEE Transactions on Vehicular Technology*, vol. 66, no. 2, pp. 952–964, 2017. DOI: 10.1109/tvt.2016.2555853.

[66] Y. Huang, A. Khajepour, M. Khazraee, and M. Bahrami, A comparative study of the energy-saving controllers for automotive air-conditioning/refrigeration systems, *Journal of Dynamic Systems, Measurement, and Control*, vol. 139, no. 1, p. 014504, 2016. DOI: 10.1115/1.4034505.

[67] H. Wang, Y. Huang, A. Khajepour, and C. Hu, Electrification of heavy-duty construction vehicles, *Synthesis Lectures on Advances in Automotive Technology*, vol. 1, no. 2, pp. 1–106, 2017. DOI: 10.2200/s00810ed1v01y201710aat001. 50

[68] M. Fayazbakhsh, F. Bagheri, and M. Bahrami, A self-adjusting method for real-time calculation of thermal loads in HVAC-R applications, *Journal of Thermal Science and Engineering Applications*, vol. 7, no. 4, p. 041012, 2015. DOI: 10.1115/1.4031018. 53

[69] X. Zeng and J. Wang, A parallel hybrid electric vehicle energy management strategy using stochastic model predictive control with road grade preview, *IEEE Transactions on Control Systems Technology*, vol. 23, no. 6, pp. 2416–2423, November 2015. DOI: 10.1109/tcst.2015.2409235. 56

[70] X. Tang, W. Yang, X. Hu, and D. Zhang, A novel simplified model for torsional vibration analysis of a series-parallel hybrid electric vehicle, *Mechanical Systems and Signal Processing*, vol. 85, pp. 329–338, 2017. DOI: 10.1016/j.ymssp.2016.08.020.

[71] X. Tang, J. Zhang, and L. Zou., Study on the torsional vibration of a hybrid electric vehicle powertrain with compound planetary power-split electric continuous variable transmission, *Proc. of the IMechE, Part C: Journal of Mechanical Engineering Science*, vol. 228, no. 17, pp. 3107–3115, 2014. DOI: 10.1177/0954406214526162. 56

[72] H. J. Ferreau, C. Kirches, A. Potschka, H. G. Bock, and M. Diehl, QpOASES: A parametric active-set algorithm for quadratic programming, *Math Program Computer*, vol. 6, no. 4, pp. 327–363, April 2014. DOI: 10.1007/s12532-014-0071-1. 56

[73] J. Liu and H. Peng, Modeling and control of a power-split hybrid vehicle, *IEEE Transactions on Control Systems Technology*, vol. 16, no. 6, pp. 1242–1251, November 2008. DOI: 10.1109/tcst.2008.919447. 59, 60

[74] N. Shidore, et al., PHEV all electric range and fuel economy in charge sustaining mode for low SOC operation of the JCS VL41M Li-ion battery using battery HIL, *Proc. of the Electric Vehicle Symposium*, vol. 23, 2007. 65

[75] B. Zhang, C. C. Mi, and M. Zhang, Charge-depleting control strategies and fuel optimization of blended-mode plug-in hybrid electric vehicles, *IEEE Transactions on Vehicular Technology*, vol. 60, no. 4, pp. 1516–1525, 2011. DOI: 10.1109/tvt.2011.2122313. 65

Authors' Biographies

YANJUN HUANG

Yanjun Huang is currently a Postdoc Fellow of Mechanical and Mechatronics Engineering with the University of Waterloo, where he received his Ph.D. in 2016. He received an M.S. in Vehicle Engineering from Jilin University, China in 2012. He is working on advanced control strategies and their real-time applications; vehicle dynamics and control; intelligent vehicle control; HVAC system modeling and control; modeling of hybrid powertrains, components sizing and power management control strategies design through concurrent optimization, and HIL testing; and variable valve actuation system for engines.

SOHEIL MOHAGHEGHI FARD

Soheil Mohagheghi Fard is a research engineer in the automotive industry. He obtained his Ph.D. in 2016 from the Department of Mechanical and Mechatronics Engineering at the University of Waterloo. His Ph.D. studies focused on development of a fuel-efficient anti-idling system for service vehicles. His research interests are: developing new powertrain technologies, power management system of hybrid and electric vehicles, and vehicle dynamics

MILAD KHAZRAEE

Milad Khazraee received his B.Sc. from K.N. Toosi University of Technology in 2010, his M.Sc. from the University of Manitoba in 2012, and his Ph.D. from the University of Waterloo in 2016, all in Mechanical Engineering. He is currently a postdoctoral fellow at the University of Waterloo. His main research interests include dynamics, system modeling, control, design optimization, power management, and hardware in the loop with focus on hybrid and electric vehicles, autonomous driving, robotics, and bio-engineering.

HONG WANG

Hong Wang is currently a research associate of Mechanical and Mechatronics Engineering with the University of Waterloo. She received her Ph.D. from the Beijing Institute of Technology in China in 2015. Her research focuses on the component sizing, modeling of hybrid powertrains, and energy management control strategies design for hybrid electric vehicles; intelligent control theory and application; and autonomous vehicles.

AMIR KHAJEPOUR

Amir Khajepour is a professor in the Department of Mechanical and Mechatronics Engineering at the University of Waterloo. He holds the Canada Research Chair in Mechatronic Vehicle Systems and senior NSERC/General Motors Industrial Research program in Holistic Vehicle Control. He has applied his expertise in several key multidisciplinary areas including system modeling and control of dynamic systems. His research has resulted in many patents and technology transfers. He is the author of more than 400 journal and conference publications as well as several books. He is a Fellow of the Engineering Institute of Canada, the American Society of Mechanical Engineers, and the Canadian Society of Mechanical Engineering.

Printed in the United States
by Baker & Taylor Publisher Services